Studies in Biotechnology Series No. 5

TRANSGENIC PLANTS

PROMISE OR DANGER

TRANSGENIC PLANTS

PROMISE OR DANGER

B. L. KAKRALYA
Department of Plant Physiology
S. K. N. College of Agriculture
Rajasthan Agricultural University, Jobner

ISHITA AHUJA
Department of Plant Breeding and Genetics
Punjab Agricultural University
Ludhiana

Agrobios (India)

Published by:
Agrobios (India)
Agro House, Behind Nasrani Cinema
Chopasani Road, Jodhpur 342 002
Phone: 91-0291-2642319, Fax: 2643993
E. mail: agrobios@sify.com

Agrobios (India)

First Published (2001)
Reprinted, 2005

9 788177 540987
Rs. 2000.00 / US$ .00
Published by:
AGROBIOS (INDIA)

Published by: Dr. Updesh Purohit for Agrobios (India), Jodhpur
Laser Typeset at: Yashee Computers, Jodhpur
Cover Design by: Shyam
Printed at: Bharat Printing Press, Jodhpur

Preface

Words and terms are frequently the root of problems associated with genetic engineering. *Transgenic* sounds very much like *eugenic* and could, therefore, be condemned by association, and *engineered* sounds more sinister than *domesticated* (Jones, 1994), though they are virtual synonyms.

Transgenic plants or Genetically Modified Plants refer to plants in which functional foreign genes have been inserted. They help understanding the mechanisms of gene expression. This could be possible only by using tissue culture and recombinant DNA techniques (efficient gene cloning methods, improved vector systems, organ specific promoters, marker genes, cloned DNA, etc.). They are useful for in following manners: (1) Agronomic traits can be improved by producing plants with increased resistance to pest, disease, stress along with increased vigor, yield. (2) Enhanced efficiency of physiological process like photosynthesis and improved nutritional qualities. (3) Transgenic plants are also used as an analytical tool to explore aspects of gene regulation. (4) Major advantage of molecular breeding is that when a particular gene has been isolated and reconstructed it can be first in model plants and later it can be used in a variety of cultivars of different crops.

Transgenic plants may be used as a means of bioremediation i.e. decontamination of polluted land, in industries (as biomass producing crops, lognin/paper, plastics, terpenoids, etc.), house hold- (clothes and ornamentals), food and drink (modification of protein, starch, oil and vitamins, allergens/toxins, beer, etc.), medical and veterinary applications (Vaccines/antigens/ antibiotics, enzymes, etc.).

The above advantages are the one face of the transgenic plants, however, their other face is little bit dangerous. The present book covers both the facts of transgenic plants with their evaluation.

Authors

Contents

2 GENETIC ENGINEERING AND TRANSGENIC PLANTS 22

3 GENE ISOLATION AND IDENTIFICATION 37

1

Biotechnology and Transgenic Plants

Opportunities for producing transgenic varieties are endless.

Organisms into which heterologous DNA (transgene) has been artificially introduced and integrated in their genomes are called transgenics.

In past, farmers have been plant breeders for ten thousand years, altering the genetic integrity of most crops without any knowledge of heredity. These farmers were the first genetic engineers (Jones, 1994), although genetics as such did not come into existence until the work of Mendel was rediscovered by De Vries (1900) and others. Using knowledge of the inheritance of traits, plant breeding assumed scientific status and during the past one hundred years has developed to allow very direct control over crop evolution for man's benefit. Until the advent of genetic engineering (GE) plant breeding was confined to making crosses within and between crop species which could occur naturally, albeit that some crosses were forced. Developing transgenic crops has become routine only within the last few years and has changed the nature of plant breeding substantially, raising a whole host of ethical considerations as a consequence.

Biotechnology is a collection of technologies that employ living systems, or compounds derived from these systems, for the production of industrial goods and services. Biotechnology is not new to the agricultural and food sectors, as man has been exploiting loving systems for the production, processing and preservation of food for centuries. What distinguishes *modern* biotechnology from the traditional plant breeding, is the emergence within the genes of recombinent DNA technology or genetic engineering.

DNA, the universal code of life, is structurally and functionally identical in all living organisms. Thus, it can be transferred between related and unrelated living organisms, and specific vectors and gene transfer systems have been developed for microbial, plant and animal applications.

THIRD WORLD NEEDS AND THE NEW BIOTECHNOLOGIES

Plant biotechnology can be said to have begun in 1983 when it was demonstrated that foreign genes could be transferred from *Agrobacterium tumefaciens*, a soil bacterium, into plant cells and stably incorporated into the plant genome and expressed (Fraley et al., 1983; Bevan et al., 1983; Chua, 1994, 1996).

World over the modern biotechnology has been revolutionized with the innovations like genetic engineering, cell fusion technology bio-process technologies and structure based molecular designs including drug development drug targeting and drug delivery system.

According to Financial Express (04-12-2000), the 2nd Biotechnology India-2000 conference concluded in Delhi with the hope that at least a number of transgenic varieties in cotton, rice, brassica, pigeonpea, moongbean, improved varieties of wheat, tissue culture regeneration for important species culture like mango, saffron, citrus, neem would soon be produced in the country in the coming years for major commercial activity.

It is estimated that in the 21st century, biotechnology based industrial based industrial revolution will surpass all other areas and may contribute between 40 to 50 per cent of the overall industrial production. The micro-propagation technology would provide high quality planting materials to the farming community. New biocontrol agents and bio-fertiliser packages would be made available to the farmers. Genes responsible for resistance to agronomically important plants, fungi, bacteria and viruses will be exploited in agriculture seeds with commercially valuable oils, proteins and starches will be developed, and such transgenic crops will be assimilated into agriculture.

The country would be in a position to fully utilize on a sustainable basis, the bio-resource particularly the medicinal and aromatic plants. Plants as bio-reactors would be able to produce large number of proteins of therapeutic value and many other important items. The recent discovery of the gene for recalcitrant species is a landmark event. The molecular biology route for development of new diagnostic kits and vaccines for major diseases, emerging and re-emerging infections would make the healthcare system much more efficient and low-cost.

Viewed in the global context, this millennium would usher in an era to conquer cancer, grow new blood vessels in the heart, block the growth of blood vessels in tumour, create new cells from stem cells and have designer crops with desired traits to give plants of nutritional value and higher productivity plants as bio-reactors. New vaccines and diagnostics would be produced through recombinant DNA technology.

With the introduction of new innovations in biotechnology in the form of novel transgenics of all types, bacterial factories to use the plant genes as source of bio-chemicals, pharmaceuticals, enrichment of genetic diversity, oral vaccines, other recombinant vaccines, tissue culture as a major commercial venture, application of DNA fingerprinting techniques especially for endangered species and many others would be available to mankind.

Basic need	Potential contribution of new biotechnologies	Dominant research of biotech industry
Conservation and improvement of diverse people's crops emphasizing hardiness, nutrition, and yield.	Crop production Tissue culture technology could support conservation and breeding objectives.	Rather than pest resistance the focus is on gene transfer for pesticide resistance, encapsulated embryos and yield improvement for major crops only.
Key concerns are durability, nutrition, and cost, Product and production should be culturally and environmentally sensitive making the best use of local resources.	Food processing Improvement of traditional fermentation methods and development of new possibilities.	Focus is on reducing or substituting raw materials and the factory production of agricultural products.
Conserve diversity and broaden breeding efforts for foraging animals to develop healthier, more efficient livestock.	Animal husbandry Vaccines and diagnostic can support these efforts and embryo transfer can help preserve diversity.	Attention is on complete control over fertility and reproduction to develop high yielding uniform, but highly vulnerable breeds and also on veterinarial packages and on use of livestock as bio-reactors for drugs.
Best way to improve health is to eliminate poverty. Following that	Health Care Biotechnologies could help with monoclonal	Emphasis is on diagnostics and clinical asays, help against

Basic need	Potential contribution of new biotechnologies	Dominant research of biotech industry
preventive health care focusing on improved sanitation, nutrition and drinking water. Next, new vaccines for tropical diseases and AIDS.	antibodies for water testing and gene technology for vaccine research and production.	infertility, production of hormones and drugs related to aging, cancer, AIDS, heart disease and organ transplants and gene therapy.

Priorities for Indian biotechnology in agriculture and the environment are:

1. Bio-fertilisers
2. New varieties
3. Nitrogen fixation
4. Quick propagation through tissue culture
5. Improvements to animal health and productivity
6. Conservation of forests and afforestation
7. Pollution recycling
8. Waste recycling

TECHNIQUES OF GENETIC ENGINEERING: PRESPECTIVES

The term genetic engineering broadly refers to genetic manipulation of cells and organs at molecular level. In the past, Mendelian genetics and conventional breeding methods have been the tools for the genetic improvement of organisms. Today efficient procedures for the successful culture of somatic cells, pollen, protoplasts and plant regeneration from a large number of plant species and artificial breeding methods in animals have been developed. Besides the protocols for organism development, availability of various restriction enzymes, DNA vector systems, transposable elements, series of promoters, marker genes and a large number of cloned DNA fragments/genes, have made genetic manipulations more precise and directed.

Such developments have further widened the scope of genetic engineering to precisely insert foreign genes into any organism from diverse biological systems (plants, animals, and microorganisms), in cloning of useful genes and to facilitate molecular marker based selection for the complex traits. Through the use of various molecular probes, characterization and introgression of specific DNA sequences have become more precise and rapid.

Genetic engineering has brought about radical changes in both protein and nucleic acid chemistry and in the use of certain novel biological systems as selection agents themselves. The analytical techniques associated with recombinant DNA have developed rapidly. DNA synthesis and sequencing can both be accomplished far more rapidly today than was 10 years ago. The tremendous need for constructing large number of specific genes created requirement of DNA sequencing and led to invention of automated sequencers to confirm the actual sequence of the construction. Gene constructs, when introduced into the recipient organism, require analyses of the organization of host genome, as well as, of gene expression. Southern analyses were developed to visualize the gene integration. Northern and Western blots were developed to assay gene expression by assaying specific RNAs and proteins, respectively. The polymerase chain reaction (PCR) technique opened new vistas including the evaluation of gene integration and synthesis of partial gene sequences. The RFLP and oligonucleotide polymorphism techniques became important tools for genetic analysis of traits at DNA level and quantitative trait loci (QTL) analysis. These supporting technologies, as well as a number of others, relevant to nucleic acid and protein chemistry, are major reasons that genetic engineering has progressed at a rapid rate.

These days, gene isolation and genetic transformation are routine in many laboratories. Transgenic plants and animals have been produced in more than 100 species. Besides, selectable marker genes including those for chloramphenicol acetyltransferase, neomycin phosphotransferase, hygromycin phosphotransferase; the genes conferring herbicide tolerance, insect resistance, virus protection, storage proteins, and environmentally regulated genes have been transferred to the transgenic plants from a wide range of living organisms. In majority of the cases, foreign genes show expression in transgenics and are stably inherited to the progeny without any detrimental effects. More interestingly, transgenic plants under field conditions have also maintained increased level of insect resistance, herbicide tolerance and higher level of virus protection.

Transgenic animals have also been created for their potential use in diverse fields. The new engineered animals are disease resistant and provide meat and milk at a lower cost of production. Besides, they serve as bioreactor producing useful drugs, vaccines, hormones, organs for transplantation and many other products benefiting mankind. Preparation of animal clone is an everyday news. Use of genetic engineering in forensic science, human diseases, controlling environmental pollution and even at industry-level (though low in frequency at present), is expanding at high pace.

The recent indications suggest that in addition to scientific probe in biochemical traits, there is a growing interest in the market place for traits with "added value" like oilseeds with greater oil content or more healthful fatty acid profiles, feed grains that have improved nutritional qualities, meat with low levels of cholesterol or even bovine milk with antibodies etc. These marketplace changes are complex and can not be easily understood as they are related to consumer demand, convenience, diverse population, health consciousness, and a more competitive world market.

In promising miraculous future benefits and applications, genetic engineering has invited public attention too. The technique of redesigning the basic blueprint of life in a deliberate manner, rouse threats of indiscriminate release of recombinant DNA-containing organisms into the environment. The risks include harmful biological impact on gene pools and ecosystems, as well as that on global economy, social values and legal structures of our existing systems. The immediate worry revolved around biohazards due to the presence of chimeric or synthetically reorganized organisms in the environment at large. It was soon apparent, that much more was at stake than immediate health and ecological imbalances. A technology that possesses the power to genetically redesign even human beings could be expected to have significant impact on ethical, moral, social, economic and legal values and structures.

Though dramatic progresses have been made in genetic engineering of crop plants, several factors still need investigation before the potential of this technology can be realized on commercial scale. Plant regeneration has been achieved in a number of plant species, but it is often limited to specific genotypes. Hence, to incorporate desirable genes into crop plants, it is essential to have better plant regeneration systems applicable to a wide range of elite germplasms. In spite of the extensive information available in molecular biology on the use of organ-specific promoters, very little is known about the mechanism of gene expression particularly how to switch the genes on and off. Moreover, experiment utilizing a large number of transgenic plants pointed out loose association between high copy number of the T-DNA and non-Mendelian inheritance of certain genes. Rearrangements and multiple insertions are also frequent events during *Agrobacterium*-mediated transformation of plant cells. Thus, there is a need to have extensive laboratory and field testing of transgenic plants and their progenies before their usefulness can be realized in crop improvement for commercial application. The availability of new cloned DNA fragment/genes would be increasingly useful in monitoring alien gene introgression into the organism. Further, isolation of new genes would accelerate our understanding of the mechanism (regulation) of gene expression and in exploiting its potential for future benefits.

In nut shell, the future of genetic engineering is bright. Close interaction between solving fundamental problems and those exploiting opportunities for developing improved germplasm with innovative properties would open vistas for efficient use of biological system for mankind.

TRANSGENIC PLANTS

Words and terms are frequently the root of problems associated with GE. *Transgenic* sounds very much like *eugenic* and could, therefore, be condemned by association, and *engineered* sounds more sinister than *domesticated* (Jones, 1994) though they are virtual synonyms.

Transgenic plants refer to plants in which functional foreign genes have been inserted. They help in understanding the mechanisms of gene expression. This could be possible only by using tissue culture and recombinant DNA techniques (efficient gene cloning methods, improved vector systems, organ specific promoters, marker genes, cloned DNA etc.). The techniques for transferring foreign genes into plant cells regenerating the transformed cells in fertile transgenic plants were established by Bevan et al, (1983) and Fraley et al, (1983). Thus, the year 1983 was a watershed for plant biotechnology. The technical break through fueled not only activities in basic research on the mechanism of plant gene regulation, but also intensive exploitation of its biotechnological application for crop improvement (Chua, 1994, 1996).

During the last 17 years, dramatic progress has been made in the application of transgenic technology for crop improvement. Because the transfer of one or two genes can be done reproducibly and routinely with this technology it is not surprising that the first generation experiments in plant biotechnology entailed the transfer and modification of single gene traits. These experiments were directed towards two commercial target areas: plant protection and quality enhancement (Chua, 1996). The subject has been reviewed and overviews (Goodman et al., 1987; Uchimiya et al. 1989; Paszkowski et al. 1989; Raffa, 1989; Cornelissen and Melchers, 1993; Chua 1994, 1996; Kaul and Nirmala, 1995; Goy and Duesing, 1995; Beachy, 1999, Tyagi, et al. 1999).

ADVANTAGES OF TRANSGENIC PLANTS

1. Agronomic traits can be improved by producing plants with increased resistance to pest, disease, stress along with increased vigor, yield.
2. Enhanced efficiency of physiological process like photosynthesis and improved nutritional qualities.
3. Transgenic plants are also used as an analytical tool to explore aspects of gene regulation.

4. Major advantage of molecular breeding is that when a particular gene has been isolated and reconstructed it can be first in model plants and later it can be used in a variety of cultivars of different crops.

Other Uses

Environment

Transgenic plants may be used as a means of bioremediation i.e. decontamination of polluted land. Transgenic *Liriodendrom tulipifera* (yellow poplar) expressed bacterial mercuric reductase gene (Rugh et al., 1988). These plants survive in high Hg levels and release elemental Hg.

Industry

Biomass/fuel: Biomass producing crops include rapidly growing tress used for fuel or products (ethanol). Transgenic plants for this aim either overexpress cellulose synthase gene (Arioli et al., 1998) or cellulases that convert cellulose to ethanol (Lebel et al., 1998).

Lignin/paper: Cellulose is raw material for paper. Lignin is separated from cellulose fibers during pulping. Transgenic plants with modified lignin (Lapierre et al., 1999) show ease of lignin solubilization and fragmentation. The only difference is the red brown coloured wood.

Plastics: High value polymeric compounds like carbonyl include non calorific sweetness like fructan produced in transgenic sugar beet (Seveneir et al., 1998). Biodegradable plastics include polyesters of 3-hydroxy acids i.e. polyhydroxy alkanoates (PHA).

Terpenoids: Aromatic qualities of essential oils is a feature of monoterpenes and sesquiterpenes belonging to biongoing to bio prenoid group of compounds. Construct of c-DNA library from oil gland secretory cells of peppermint has allowed menthol. Another way increasing essential oil synthesis is by increasing density of mint glandular trichomes.

House hold- (clothes and ornamentals)

Transgenic approach on cotton include introduction of pigment compounds and synthesis of plastics like PHB in same cells. Coloured fibres avoid dying process whereas plasticity lowers thermal conductance. Genetic manipulation of ornamentals can produce coloured flowers that do not occur naturally. Vase life is increased by generating the antisense down regulation of an ethylene biosynthesis gene.

Food and Drink

Protein, sugars, starch, oil and vitamins: Proteins can be modified either by changing the amino acids in a protein or by adding entirely new protein. Another application is expression of human milk proteins in edible plants. For example, gene for beta casein protein can be introduced in potato under the control of an amino auxin inducible promoter. Similarly, lactoferrin (Salmon et al., 1998) proteins have been expressed in transgenic tobacco. It has antibacterial, anti fungal and antiviral properties.

Degree of branching in carbohydrates can be altered (Heyer et al., 1999) and degree of saturation and chain length of oils can be altered (Murphy, 1999) to produce transgenic plants useful for human and industry.

To overcome vitamin A deficiency psy gene encoding enzyme phytoene synthase can be introduced in plants deficient in vitamin A, production of carotenoid (vit A) is enhanced. Similarly, content of micronutrients can be increased. Endosperm of transgenic rice had high levels of Fe when gene encoding soyabean ferritin (an iron storage protein) was introduced and expressed under control of rice seed storage protein promoter (Goto et al., 1999).

Allergens/Toxins: Allergens can be successfully reduced in transgenic plants. Example, level of allergenic proteins (eg. AraHl) in peanut can be reduced. Similarly, caffiene level can be reduced by antisense down regulation.

Beer: Transgenic barely is generated involving the modification of the glucanase enzyme as means of improving the efficiency of brewing process. (Jensen et al., 1996). Introduction of leuco anthocyanidin reductase gene improves malt quality.

Medical and veterinary applications

Vaccines/antigenes/antibiotics

This area include production of plant based vaccines and other medically related compounds. Transgenic potato have been successful in expressing human insulin (auto antigen) either alone or as a protein in which insulin was linked to c-terminus of cholera toxin B sub-unit (CTB).

Enzymes

Production of active enzymes is another area of interest. "World's most expensive drug" is the human lysosomal enzyme glucocerebrosidase, the only treatment for patients suffering from Gaucher's disease. It is purified from human placentae. Thus, it's production in plants is an attractive option.

TRANSGENIC ARE NOT TERMINATOR CROPS

The most debated topic in agriculture and biotechnology is genetically modified (GM) crops, that is, transgenics. Genetic engineering (GE) techniques allow the transfer of a gene of interest from one species to another. It is nothing but the extension of conventional plant breeding. This technology helps the breeders and molecular biologists to introduce the only gene of interest with more selective modification and represents a significant advance.

Terminator technology is a technique of genetic engineering by which the plant is altered or modified in such a way that it produces sterile seeds, which will not germinate in the next generation. Normally, seeds have the ability to germinate generation after generation. In terminator seed, a toxic gene becomes active and is forced to die after one generation of survival. In terminator technology, three gene constructs are introduced through GE techniques to express the desired character, sterility.

The three gene constructs or stretches of DNA are referred as gene X (lethal gene), Y and Z. This entire system is under the control of an external inducer, that is, tetracycline. If only the lethal gene is inserted, then the first generation itself will not germinate. Initially, the lethal gene is prevented from being expressed for seed multiplication and thereafter, gene Y helps to excise out the 'blocking sequence' to make the gene X (ledder the control of a different gene, Z. The lethal gene becomes functional only at a later stage of seed development and allows the gene to produce toxin that kills the embryo. Ultimately, the seed harvested by the farmer lacks germination ability but according to patent claim, the seed or fruit can be used as food.

Terminator technology is developed to serve following two main purposes:

1. To force the farmers to buy seed each year for raising the next crop.
2. To prevent the piracy of the introduce special genes (that is biotic and abiotic resistance or nutritional characteristics) by researchers and multinational companies.

The Hypothetical Risks

1. *How safe are these produced seeds from the terminator seed for human consumption?*
2. *How safe to the environment as the activation of terminator gene requires pre-soaking of the seed in tetracycline?*
3. *Will these seeds respond differentially to environmental changes or have a devastating effect on bio-diversity?*

The beneficial side is that this technology has the ability to prevent escape of transgenes to bio-diversity.

Biotechnology is used to harness the problems that are difficult to overcome by traditional plant breeding techniques. GE techniques are primarily applied to resist biotic and abiotic stress, higher productivity, better seed and nutritional quality. This technology is helpful in resistance breeding, introducing gene, irrespective of their presence in nature. 'Bt' gene from the soil, *bacterium Bacillus thuringiensis* is introduced to crop plants for controlling lepidopteron insects in many crop plants.

In cotton, Bollgard a transgenic variety resistant to cotton genic variety resistant to cotton bollworm, is already in commercial cultivation and gained popularity. The herbicides are used to kill weeds but, unfortunately, they kill crop plants too. Therefore, the transgenic crop plant tolerant to herbicide are an important requirement to control weeds through herbicides. Development of bio-pesticides is an emerging area of molecular breeding, with positive contribution to integrated pest management. GE techniques are used to improve the virulence of natural enemies of pests.

The major component of a human diet is oil and fat, in which saturated fatty acid compounds are present, which are hazardous to health. GE techniques have proved their ability to reduce the level of saturated fatty acid components in soybean and castor. Balancing of human nutrition, may be carried out by trangenics to supply balanced source of amino acid. Curing human disease is becoming possible through the use of transgenic plants. Recently, it was reported that transgenic potato and tobacco can cure cholera. A transgenic potato can also prevent diabetes. European researchers developed a vitamin rich (carotenoids) tomato, which eventually, could prevent heart and cancer diseases.

Biotechnology has the potential to help society solve serious problems, but the new technology has to be handled very cautiously to protect the flora and fauna from unintended consequences. The fear is that the transgenic crops will become weeds and that novel genes may be transferred to wild populations, leading to *superweeds*. This is not that easy, as it depends on the nature of pollination and many other factors. In case of self-pollinated crop, the extent is less than five per cent, whereas in cross-pollinated crop it is up to 50 per cent and for wild and weedy relatives, the gene flow is very less (less than 1 per cent). The effect of gene flow in case weedy species is not dangerous enough to cause extinction of biodiversity. Hence, the fear of loss of entire bio-diversity needs further critical examination.

Though the inception of genetic engineering in agriculture has to faced scientific and technology challenges, economic, regulatory and social concerns, it will emerge as an important tool in solving the problems of food production healthcare, bio-diversity and environment conservation. The present world population is around billion by 2020, it may becomes billion and out of which 85 per cent would be from the developing countries. It is time to increase India's agricultural productivity which is still far below the world average in most food crops.

The **Green Revolution** was possible due to the genetic manipulation of the plants by plant breeding. After the Green Revolution, most crops are suffering from yield plateauing and further boost may be possible from molecular breeding. To meet the challenges ahead, funds must be diverted towards new technologies to improve agricultural productivity Biotechnology may be able to solve some of the problems associated with lower productivity, and could play a major role in agricultural production.

A misconception still prevails among farmers about transgenic and terminator technology. They confuse one for the other and think that any transgenic crop will lead to sterility in the next generation crop. In both cases, the product is developed by using the same techniques of genetic engineering but the products are developed for different purposes.

A survey conducted among Indian farmers shows they are highly innovative and interested in adopting any new technology. However, it should be made available to them. A misconception still prevails among farmers about transgenic and terminator technology. They confuse one for the other and think that any transgenic crop will lead to sterility in the next generation crop. In both cases, the product is developed by using the same techniques of genetic engineering, but the products are developed for different purposes.

The governments of developing countries have identified biotechnology as a promising and key research area with widespread application in diverse field. The developed nations benefit from biotech products because, of both private and public sector collaboration. Though there are long term risks and hazards associated with it, this new technology should be used to complement, not replace, traditional plant breeding methods to enhance productivity and quality.

Ban on Terminator Seeds to Stay

According to Union Agriculture Minister, Mr. Nitish Kumar, (The Hindu, December 1, 2000), said, *No terminator seed has been or will be allowed in the country.* He was replying to Mr. Eduardo Faleiro, he said the Government had not given permission to transnational companies to import terminator seeds. *This was not being considered by the Government.*

CLEAN CHIT TO GENE CROPS

The Government of India has given a clean chit to transgenic crops, saying studies indicating that they were safe from environmental and health hazards (Business Line, December 13, 2000). An official release quoted studies on food safety and environment impact conducted by major research institutes across the country to say that work on transgenic plants would continue for improving productivity, nutritional and environmental security with stringent bio-safety regulations. Studies on food safety were conducted by the Industrial Toxicological Research Centre, Lucknow, and Sriram institute here, and on environment impact by Acharya N G Ranga Agricultural University, Punjab Rao Krishi Vidhyapeth and Marathwada Agricultural University.

GENOMICS AND TRANSGENIC PLANTS

The term genomics was coined by Thomas Roderick (1986) to describe the scientific discipline of mapping, sequencing, and analyzing genomes and to provide a name for a new journal – *Genomics*. Genomics is now undergoing, however, a transition or expansion from mapping and sequencing of genomes to an emphasis on genome function.

The coded genetic information of organisms goes by the new buzzword genomics. Useful organisms are now being uncovered, *en masse,* in genome organization, expression, and interaction, rather than in terms of individual genes. The recent advances in genetic manipulation were possible as it is now possible to define the gene with respect to its function.

Today, any gene of interest from any species may be identified, isolated and incorporated across species barriers. The gene has great impact on all aspects of our lives – from food production to health care, from society to politics. Recently, genes and genomics became the most discussed terms in several areas of Government and law, as they became important items of intellectual property due to the application of IPR on biological organisms.

Plants are a key source of energy for the survival and evolution of the animal kingdom. Over the last few decades, plant genomics has been gaining importance for the improvement of plant characteristics. Genomics is also important for phylogenetic analysis and adds new dimensions to evolutionary theories, which is likely to revolutionise plant breeding research.

The major revolution in the field of genomics in eukaryotes is possible due to the availability of recombinant DNA and PCR technology from the 1970s and 1980s, respectively. Though genomics is still in its infancy, it will increase the understanding of basic plant biology as well as the exploitation of genomic information for crop improvement.

Genomics is the study of how genes and genetic information are organized within the genome, and how this organization determines their function in a given organism-plant or animal. Genomics covers the identification, mapping and sequencing of the gene. The objective of genome research in any species is to sequence the entire genome and to decipher functions of all the different coding and non-coding sequences.

Today's advances in genetics, miniaturization, computer-aided techniques and sophisticated data management have increased the possibility of solving the inherent problems within plant and animal systems. The management of the vast pool of genetic information generated by genomics research into easily accessible form has grown into a new branch of science-bio-informatics.

Bioinformatics helps in converting this data into an understanding at the gene function, which determines the scope for the successful application of genetics in humankind. The sequencing of an entire genome is time-consuming, and that is why genome technology is used to develop a cost-effective and rapid way to describe the entire genome of an organism approach (that is, *expressed sequence tags* or ESTs).

The development of ESTs for individual genomes is also arduous, and has led to the development of a new tool – DNA chip technology. This technology helps parallel data acquisition for millions of DNA sequences through automation at a very fast rate using computer devices.

Genomics research is divided into two groups based on the information generated. Structural genomics – a study of the structure of all proteins encoded in the DNA sequences of an organism; and functional genomics – the study of the function of all specific gene sequences, effects and their expressions in time and space represented by the DNA sequences. How and why a given gene behaves in a certain way under specific conditions. Functional genomics also includes research on the protein function – proteomics – of an organism.

Structural genomics represent an initial phase of genome analysis and involves construction of high resolution genetic, physical and transcript map of the organism. The ultimate physical map of an organism in its complete DNA sequence. Functional genomics refers to the development and application of global (genome wide or system wide) experimental approaches to assess gene function by making use of the information and reagents provided by structural genomics (Hieter and Bogurski, 1997). It is characterized by large scale experimental methodologies combined with statistical and compartmantional analysis of results.

In crop science, genomics research has been promoted using *Arabiodopsis* as a model plant for the dicot family: this has the smallest plant genome of 140 million base pairs, and is used extensively for genetic experiments. It is a weed found all over the world and grown easily. This entire genome sequence will be available by this year end, and will provide the blueprint of the basic set of genes required for growth and maintenance of the plant. Similarly, in monocot families, rice (*Oryza sativa*) has been selected as a model plant, as the genomes is relatively smaller than other cereals but three longer than that of *Arabiodopsis,* and it is also the most world's important food crop.

Genomics will also help in evolutionary studies. By comparing the genomics of lower and higher organisms, we can make out sequences, which can be conserved (hence are associated with life basic processes) and which have been deleted or evolved (such sequences will be organism specific) during evolution. For example, sequences related to intercellular communication are expected to be present in *C. elegans* but not in yeast (because it is single celled) and this is the case in reality.

Genomics and Agricultural Development

From the perspective of genomics companies, agriculture is the area with the most potential. Genomics in agriculture is related with the identification and sequencing of the genes of all major food crops, and the patenting of these newly-discovered genes. The challenge is in the next phase, which includes the right use of this information across species and generic barriers, that is, genetic engineering.

In the initial stages, transgenic crops were developed for straight forward characters, but with the inception of genomics research the strategy is to develop *designer plants* with most of the multi-genic characteristics. Genomics research can take up the challenge of coordinating the expression of several genes simultaneously to solve critical agricultural problems that are difficult to address through conventional plant breeding.

The earlier evolutionary theories were mostly based on geographical variations and morphological differences within the organisms. However, genome analysis provides detailed genetic information of the natural populations, which promise a better understanding of the evolution and phylogenetic relationships of the species. Further, genomics provides more precise information to understand the evolution from wild progenitors, better taxonomic classification, and the pattern of genetic variation within a species.

In the 21st century, there is an urgent need to increase food production to sustain the burgeoning population with no scope for the dramatic expansion of cultivable lands.

The increase in food supply is possible by reducing the losses caused by pests and diseases, changing environments and post-harvest losses. Modern genetic technology, as a new and revolutionary tool, can help in genetic improvement through transgenics by incorporating the desired gene (s) identified by genome analysis.

Genomics is the logical evolution of the electronic age for new developments in molecular information with a potential impact on agricultural research. Scientific progress strongly depends on the resource allocation and, therefore, plant genomics will certainly make vital contributions to our understanding of the problems of agricultural production.

India is committed to the WTO provisions on worldwide patent standards by 2005, and the situation will draw multinationals which will want to tap its strengths and huge scientific talent base. The sheer population pressure of the developing world is demanding a higher priority for transformation of agriculture and health-care to fulfill the demands. In the future, there are great opportunities for India to play increasingly important role in scientific activities that will benefit both the country and the world at large.

There is an urgent need to explore ways of increasing food production to sustain the burgeoning population, and to reduce the losses caused by pests, diseases and post-harvest losses. Modern genetic technology can bring about improvements through transgenics by incorporating into crops the desired genes identified by genome analysis.

RISE IN GM CROP PLANTED AREA

Planting of genetically modified (GM) crops worldwide has considerably expanded over the last five years, and during this year the area under cultivation of GM crops stands at 44.2 million hectares, up 11 per cent from the previous year's 39.9 million hectares, according to a communication from Monsanto Holdings Ltd. (Business Line, 23-12-2000). These data are based on *Global review of commercialized transgenic crops: 2000* complied by the International Service for Acquisition of Agri-biotech ((ISAAA). It is an independent agency monitoring the use of bio-tech crops globally, the communication said currently there were 13 countries spread across six continents. Of the 13, eight were industrialized countries and the rest developing nations.

Since 1996 when 1.7 million hectares under GM crops, there had been a 25-fold increase in area in the last five years. This adoption rate of bio-tech crops was the highest for any new technology by agriculture industry standards, ISAAA said.

Argentina was on the fast track for GM crops with a reported increase in area of 3.3 million hectares for 2000, reflecting an increase for modified soybean and corn and a modest increase in cotton. In the US, there was a net increase of 1.6 million hectare area in soybean, cotton and canola. South Africa increased plantings by one lakh hectares for transgenic corn and cotton; Australia by 1.5 lakh hectares for cotton; and China by two lakh hectares for Bt cotton, Monsanto said.

Interestingly, four countries US with 30.3 million hectares; Argentina with three million hectares; and China with 0.5 million hectare – accounted for 99 per cent of global transgenic crops acreage in 2000.

Specifically, a significant 43 per cent increase in global area under transgenic cotton had been witnessed in 2000 with the hectarage now estimated at 5.3 million hectares. The most significant increase was reported from the US where the percentage of GM cotton went up from 55-72 per cent. About 10 per cent of China's cotton crop area is now transgenic.

Globally, demand for herbicide tolerant crops was high followed by insect-resistant crops. During this year, herbicide tolerance deployed in soybean, corn and cotton occupied 74 per cent out of an area of 44.2 million hectares, with Bt insect –resistant crops taking up 19 per cent and stacked genes for herbicide tolerance and insect-resistance in cotton and corn according for seven per cent.

GM CROPS : GLOBAL AREA COVERAGE

The total global area coverage of genetically modified crops has increased from 1.7 million hectare in 1996 to 44.2 million hectare in the year 2000 indicating a 25-fold increase in a span of five years, according to Global Review of Commercialised Transgenic Crops: 2000 complied by the International Service for Acquisition of Agri-Biotech Applications (ISAAA).

The number of countries cultivating GMC has gone up from six in 1996 to 13 in the year 2000. Eight of them are from industrialized nations and five from developing countries. The demand for GMC from farmers in developing countries is surging because of more convenient and flexible crop management, higher productivity and safer environment through decreased use of conventional pesticides.

The study stated that earlier up to 85 per cent of GMC were grown in the industrialized countries. But with developing countries fast opting for these crops, their share of cultivation rose from 14 per cent in 1997 to 16 per cent in 1998 to 18 per cent in 1999 and 24 per cent in 2000. Now developing countries account for a quarter of the total global area coverage under GM crops of 44.2 million hectare. In fact, of the 4.3 million hectare increase in global coverage in 2000, 3.6 million hectare (equivalent to 84 per cent) was in developing countries as compared with 16 per cent (0.7 million hectare) in the industrialized countries.

Thus, the coverage of transgenic crops in developing countries grew by 51 per cent in 2000, from 7.1 million hectare to 10.7 million hectare, compared with a two per cent growth in industrialized countries where hectarage increased from 32.8 million hectare in 1999 to 33.5 million hectare in 2000.

The study indicates that since 1996 a cumulative total of over 125 million hectare of GM crops have been cultivated globally.

Agrentina reported an increase of 3.3 million hectare for the year 2000 in GM soybean and corn and a modest increase in cotton. In USA, there was a net increase of 1.6 million hectare in soybean, cotton and canola (rapeseed).

GM CROPS ARE SAFE

The studies conducted by the government indicate that transgenic crops are safe from environmental and health angle, according to an official press release (Tribune, December 13, 2000). The Lucknow-based Industrial Toxicological Research Centre and the Sriram Institute, Delhi had carried out the food safety evaluation of transgenic crops, while the environmental impact studies were done among others by Acharya NG Ranga Agricultural University, Punjab Rao Krishi Vidhyapeeth and Marathwada Agricultural University. Besides this, the monitoring-cum-evaluation committee of the department of biotechnology, comprising agricultural experts, had also made extensive field investigations.

The study stated that the work on transgenic plants would continue for improving productivity, ensuring nutritional and environmental security. Already the stringent biosafety regulations are in place. The potential of biotechnology for enhancing crop productivity and quality improvement of crops is recognized world-over.

In the meantime, as result of the integrated pest management, there has been a significant reduction in the consumption of pesticides, from over 61,350 tonne to about 46,200 tonne in the last five years.

The study conducted by the scientists is in response to the fears expressed by the general public about the genetically modified crops and foods. Many NGOs have alleged that hazardous genetically modified (GM) foods are entering the country untested and unchecked. In the time of Orissa cyclone, many such GM foods were distributed to the people by foreign agencies.

India has represented before the Codex Ali mentarius Commission and has demanded mandaroty labellings of all genetically modified foods, seeds and planting materials, so that the consumers can have their right choice. The Union agriculture ministry is planning amendments to the Seeds Act to restrict the imports of hazardous genetically modified seeds.

The Union health ministry is also planning modifications in the Prevention of Food Adulteration (PFA) Act for installing infrastructure for testing of these GM foods at all points of entry, in collaboration with the Customs department and department of bio-technology.

This measure is likely to make the importers primarily responsible, in lieu of traders for any hazardous consequences arising out of consumption of GM foods.

The study stated that the work on transgenic plants would continue for improving productivity, ensuring nutritional and environmental security.

PUBLIC ACCEPTANCE OF GENETIC MODIFICATION

As has been alluded to previously, in terms of biotechnology research, one of the most important economic considerations separating forestry from agriculture is that trees require substantially more time for proper evaluation under both laboratory and field conditions. Rapid generation turnover in crop species has allowed genetic modification technology to develop quickly. Although this has been desirable from an economic point of view, the ecological ramifications have been far more difficult to study, as complex experiments must be carried out over several years. Public and government acceptance of GM plants is now probably as dependent on biological risk assessment and risk issue management (Leiss, 1999) as it is on technical development or economics alone.

For GM trees, there is clearly a need to provide scientifically sound, neutral and intelligible information to the public, so that when and if GM trees become commercially available, informed decisions can be made about their release into the environment. Foresters are generally very aware of the need for public involvement in many other areas of forest management (e.g. chemical applications in forests, game management, felling patterns and plants, allowable annual harvesting levels), so decisions concerning GM trees may simply be an additional issue to be managed in the process. (adopted from Unasylva, FAO, Vol. 52: 2001/1).

CAN GENETICALLY MODIFIED TREES BE SAFELY DEPLOYED IN THE ENVIRONMENT?

The main ecological concerns with GM plants is the issue of gene exchange with wild populations – an issue already being considered in forestry with regard to seed orchard seed from conventional breeding programmes. In most of the situations for which GM trees will be considered, the plantation species will be an exoitic one, so gene exchange would not be a factor; but in cases where it could occur, reduced flowering or sterility is likely to be a basic requirement. Much has been and will continue to be learned from agricultural research on transgene flow from GM crops to their wild relatives. Investigation into these specific questions are under way in forestry (DiFazio *et al.,* 1999).

Although exotic plantation species may have limited risk of cross –breeding *per se,* the release into the environment of a GMO in most cases poses less of an ecological risk than the introduction of an exotic species. The introduction of an exotic species is a release into the environment of an entirely different genetic constitution, rather than the introduction of a single gene into a native species. The current regulations of the Forest Stewardship Council (FSC, 1999) prohibit the use of GM trees, but suggest that *the use of exotic species shall be carefully controlled and actively monitored to avoid adverse ecological impacts.* This recommendation largely runs contrary to the argument made here that the biological risks and economic realities of all forest management options need to be evaluated objectively. On the other hand, a large genetic change made to the overall fitness of a native species and released without adequate consideration of local environmental risks is not appropriate either.

Clearly, potential risks must be balanced with benefits in all types of improved forest trees, considering the deployment schemes that will be used in different locations and at different times. For example, clonal varieties of fruit trees have been used in large continuous orchards for hundreds of years, and this deployment and management approach is considered by most people to be ecologically and economically acceptable, as it presents no unmanageable biotic or abiotic challenges. In plantation forestry for wood production, clonal blocks may be appropriate in some situations, such as poplar plantations, while pure or family-mix planting patterns are likely to be appropriate for most other forestry situations.

In any event, deployment strategies must be signed to minimize the risks of economic losses of the stand as well as future biological losses, such as the development of resistance in pest and disease populations (Roberds and Bishir, 1997). At a minimum, therefore, a few dozen well-tested GM clones will be necessary to meet the typical production and diversity requirements necessary even in intensive forestry programmes.

To help address many of these concerns, a number of countries have developed regulations and restrictions specifying the requirements of confined field testing that are needed before commercial release (OECD, 2000). These requirements, which are necessary to reduce biological and economic risks, will undoubtedly continue to evolve. In addition to national laws and regulations, there are also broader international agreements on biosafety, e.g. the Cartagena Protocol on Biosafety (Convention on Biological Diversity, 2000). (adopted from Unasylva, FAO, Vol. 52: 2001/1).

2

Genetic Engineering and Transgenic Plants

Genetic engineering has unlimited scope for application, which include technologies that cater to human health and prosperity. The goal of plant, animal and microbial genetic engineer is to improve the quality and yield of products. In this chapter, various applications have been fragmented into three major subdivisions viz. transgenic plants, transgenic animals and miscellaneous applications.

The term 'transgenic' is used to describe plants and or animals which have had DNA introduced into them by means other than transfer of DNA from a sperm cell to an egg. The DNA transferred by these non-sexual processes may be DNA from same species, from other plant or animal species or even from a prokaryote. Today many transgenic plants as well as animals have been engineered by the scientists with special beneficial properties; few of them are discussed in this chapter.

TRANSGENIC PLANTS

Up till 1980s, production of transgenic plants was restricted mainly to tobacco, petunia or tomato of the family solanaceae. However, now in more than 40 dicot species transgenic plants have been successfully produced (Table 2.1). The production of transgenic plants has now been extended to several monocotyledons like wheat, maize, rice and oats.

Table 2.1: A list of higher plants were transgenic plants have been produced, using different methods.

I. Herbaceous dicotyledons

1. *Actinidia* sp. (kiwi)

2. *Apium graveolens* (celery)
3. *Arabidopsis thaliana*
4. *Armoracia* sp. (horse radish)
5. *Beta vulgaris* (sugarbeet)
6. *Brassica napus* (oilseed rape: canola)
7. Brassica oleracea (cauliflower)
8. *Brassica oleracea* var. capitata (cabbage)
9. *Brassica rapa* (syn. *B. campestris*)
10. *Carica papaya* (papaya)
11. *Chrysanthemum* sp. (chrysanthemum)
12. *Cichorium intybus* (chicory)
13. *Cucumis melo* (muskmelon)
14. *Cucumis sativus* (cucumber)
15. *Daucus carota* (carrot)
16. *Dianthus caryophyllus* (carnation)
17. *Digitalis purpurea* (foxglove)
18. *Fragaria* sp. (strawberry)
19. *Glycine max* (soybean)
20. *Glycyrrhiza glabra* (licorice)
21. *Gossypium hirsutum* (cotton)
22. *Helianthus annuus* (sunflower)
23. *Ipomoea batatas* (sweet potato)
24. *Ipomoea purpurea* (morning glory)
25. *Lactuca sativa* (lettuce)
26. *Linum usitatissimum* (flax)
27. *Lotus corniculatum* (lotus)
28. *Lycopersicon esculentum* (tomato)
29. *Medicago varia*
30. *Medicago sativa* (alfalfa)
31. *Nicotiana plumbaginifolia* (wild tobacco)
32. *Nicotiana tabacum* (tobacco)
33. *Petunia hybrida* (petunia)
34. *Pisum satium* (pea)
35. *Rosa sp.* (rose)
36. *Solanum melongena* (eggplant)
37. *Solanum tuberosum* (potato)
38. *Vaccinium macrocarpon* (crtanberry)

39. *Vigna aconitifolia*
40. *Vitis vinifera* (grape)

II. Woody dicotyledons

41. *Populus sp.* (poplar)
42. *Malus sylvestris* (apple)
43. *Pyrus communis* (pear)
44. *Azadirachta indica* (neem)
45. *Juglans regia* (walnet)

III. Monocotyledons

41. *Asparagus sp.* (asparagus)
42. *Dactylis glomerata* (orchard grape)
43. *Secale cereale* (rye)
44. *Oryza sativa* (rice)
45. *Triticum aestivum* (wheat)
46. *Zea mays* (corn)
47. *Avena sativa* (oats)
48. *Festuca arundinacea* (tall fescue)

IV. Gymmosperms

41. *Picea glauca* (white spruce)

ENGINEERED RESISTANCE AGAINST HERBICIDE

The development of crop plants that are tolerant to herbicide is an important approach to control weeds. Tolerance to herbicides has been one of the officiated goals of the plant genetic engineer for several reasons. First, there are several economic advantages for farmers to have crop engineered to tolerate herbicides. Second, the biochemical pathways inhibited by herbicides have been well characterized and strategies to achieve resistance are relatively simple to devise. Third, resistance to herbicides can often be achieved by the transfer and expression of a single gene in a new host plant. Finally, resistance to certain herbicides may allow the farmer to adopt the use of some of the newer herbicides, which are considered as being environmentally safe.

Most of the modern herbicides interfere with amino acid biosynthesis in plants (Table 2.2). Once the mode of action of a herbicide is established, the transfer of resistance to the herbicide can be done by two different approaches:

1. The first approach is modification of the target side.
2. Detoxification of herbicide

Transgenic plants carrying the genes for herbicide tolerance has been produced in tobacco, petunia, tomato, potato, Populus etc.

Modification of the target

This approach has been made for developing resistance against atleast three herbicides viz. glyphosate, sulphonyl urea and imidazolinones.

Table 2.2: Mechanisms of action of different herbicides and basis of achieving resistance against them in transgenic plants.

Active principle of herbicide	Inhibited pathway	Target product	Use	Basis of resistance
Amino acid biosynthesis inhibitors				
Glyphosate (Roundup)	Aromatic amino acid biosynthesis	EPSPS	Broad spectrum	Over-expression of EPSPS gene; bacterial *aroA* gene
Sulphonylurea and Imidazolinones	Branched chain amino acids	ALS	Selected crops	Mutant ALS gene
Phosphinothricin (Basta)	Glutamine	GS	Broad spectrum	Gene amplification; bar gene; detoxification
Photosynthesis inhibitors				
Atrazine (Lasso)	Photosystem II	Qb (32 kDa protein)	Selected crops	Mutant *PsbA* gene; GST gene; detoxification
Bromoxynil (Buctril)	Photosynthesis	-	Selected crops	*bxn* gene; detoxification

Resistance against glyphosate

Glyphosate, an active component of round up herbicides, is used as non-selective post-emergence herbicides. It is suggested that glyphosate inhibits aromatic amino acid biosynthesis by competitive inhibitor of enzyme 5-enol-pyruvyl shikimate-3 phosphate synthase [EPSPS]. EPSPS catalyzes the synthesis of phenylalanine, tyrosine and tryptophan. The aro A gene which codes EPSPS has been isolated and sequenced from *Escherichia coli*, *Salmonella typhimurium* and *Arabidopsis thaliana*.

There are two possible methods for producing glyphosate resistant plants by genetic engineering:

(a) transfer of a glyphosate sensitive EPSPS under the control of a powerful promoter causing over expression of the protein and/or

(b) transfer of gene which codes a mutated, glyphosate resistant EPSPS. cDNA of the EPSPS gene with 35 S cauliflower mosaic virus promoter was constructed and the transformed petunia cell lines with this construct exhibited 40-fold increase in EPSPS activity.

An alternative method is insertion and expression of mutated bacterial aro A gene coding for glyophosate resistant EPSPS. The source of the aro A gene was a mutagenized strain of *S. typhimurium* or *E.coli* and was transferred to tomato and/or tobacco. Transgenic tobacco plants, showing tolerance to commercial levels of glyphosate are field tested in USA.

Resistance against sulfonylurea and imidazolinone herbicides

The herbicidal action of sulfonylureas and imidazolinones is based on their ability to inhibit non-competitively an enzyme of branched-chain amino acid biosynthesis, acetolactate synthase (ALS). ALS is the first enzyme in the biosynthetic chain resulting in the synthesis of the branched chain amino acids valine, leucine and isoleucine.

Mutated strains resistant to sulfonylureas have been obtained from bacteria, fungi and plant cell cultures. The resistance of all these mutants can be attributed to possession of mutated forms of the ALS. Transgenic tobacco plants expressing a mutant ALS gene from tobacco or *Arabidopsis*, were produced that were tolerant to sulfonylurea herbicides.

Resistance against Phosphinothricin

L-Phosphinothricin (PPT) is a naturally occurring amino acid with herbicidal activity and is active ingredient of herbicide 'Basta'. PPT is a potent inhibitor of glutamine synthase and thus causes rapid increase in ammonia concentration in plants. This leads to death of plant cell. 'Bar' gene from *Streptomyces hygroscopicus* encodes phosphinothricin acetyl transferase enzyme that acetylates the free-NH_2 groups of phophinothricin and inactivates the herbicide Basta. This 'bar' gene with CaMV 35S promoter has been transferred to tobacco, tomato and potato plants using *Agrobacterium* Ti plasmid vector system. The transgenic plants have been tested for 'Basta' resistance.

Resistance against Atrazine

Atrazine is another herbicide, which inhibits photosynthesis. A 32kDa protein, Q_B is responsible for herbicide binding and is encoded by psb A gene. The DNA sequences of psb A are highly conserved in different plants.

More than 20 weed species have developed resistance to atrazine herbicides, for example, *Amaranthus hybridus*.

Detoxification or degradation of herbicide

This approach is based on selective use of herbicide, which will kill the weeds and not the crop. A number of detoxifying enzymes have been identified in plants as well as microbes. Few examples are given here: *(i) glutathione-S transferase* or GST (in maize and other plants): It detoxifies the herbicide atrazine; *(ii) nitrilase*: It is coded by gene bxn in *Klebsiella penumoniae*, and it detoxfies the herbicide bromoxynil, and *(iii) phosphinothricin acetyl transferase* or PAT (coded by *bar* gene in *Streptomyces* and transgenic plants using the *bxn* gene from *Streptomyces* spp).: It detoxifies the herbicide PPT (L-phosphinothricin).

Transgenic tomato plants using the *bxn* gene from *Klebsiella ozaenae* and *bar gene* from *Streptomyces,* as well as, transgenic potato, oilseed rape (*Brassica napus*) and sugarbeet using *bar* gene from *Streptomyces* have been obtained and were found to be herbicide resistant. Other crops for engineered herbicide tolerance include soybean, cotton and corn.

ENGINEERED RESISTANCE AGAINST PEST

Pests and insects cause severe damage to our crops and thus use of insectides and pesticides is a common measure. With genetic engineering, crops can now be protected with insects without applying any synthetic chemical. More than 30 crop species have been successfully transformed including corn, cotton, soybean, and rice with insecticidal properties.

For transgenic plant production with pest resistance properties, two major approaches have been developed: (a) the introduction of protein delta endotoxins from Bt and, (b) introduction of protease inhibitors in plants.

Bacillus thuringiensis endotoxins

The entomocidal bacterium *Bacillus thuringiensis*, upon sporulation normally produces a parasporal crystalline toxin. When ingested by a susceptible insect, a combination of the high pH and proteinases of the insects midgut solubilizes the crystal and yields active toxin. The effects of the toxin occur within minutes of indigestion, leading to disruption of insect's midgut cells. Bt toxin activity has been found against many species of insects within the orders of Lepidoptera, Diptera and Coleoptera.

These delta endotoxins are very specific in their action and are safe insecticides, but their use is limited due to high production cost and due to instability of crystal proteins when exposed in the field.

The above toxin gene (*ht2*) from *B. thuringiensis* has been isolated and used for *Agrobacterium* plasmid mediated transformation of tobacco, cotton potato and tomato plants. These transgenic plants were resistant to the feeding damage of *Manducta sexta* larvae (tobacco horn worm), a pest of tobacco. Experiments of feeding the larvae of *M. sexta* on leaves of these transgenic plants showed 75%-100% mortality of the larvae, while the control plants carrying no transgenes were severely damaged.

Pseudomonas flourescens that Secretes Bt Toxin

The cystalline protein toxin produced by *Bacillus thuringiensis* var. *israelensis* is toxic to the larval stage of several mosquitoes and black fly species. Normally this toxin is produced by fermentations of *B. thuringiensis* and the toxin is sprayed onto crops. Scientists have produced genetically engineered *Pseudomonas fluorescens* living in the roots of corn plants. This bacterium can secrete Bt-toxin to kill black cut worm.

Field tests using transgenic insect resistant plants were also conducted with tobacco and tomato and may be used for commercial cultivation in the near future.

Table 2.3: Insecticidal efficacy of CpTi in artificial diets and in transgenic plants

Insects killed by CpTi in artificial diets	Insects, against whom resistance noticed in transgenic tobacco plants (with CpTi)
Lepidoptera	
Heliothis virescens	*Helothis virescens*
Heliothis zea	*H. zea*
Spodoptera littoralis	*Spodoptera littoralis*
Chilo partellus	*Manducta Sexta*
	Autographa gamma
Coleoptera	
Callosobrichus maculatus	
Anthonomus grandis	
Diabrotica undecimpunctata	these insects do not attack control tobacco plants
Tribolium confusum	
Costelytra zealandica	
Orthoptera	
Locusta migratoria	

Protease inhibitors

Protease inhibitors are proteins with antimetabolic activity against wide range of insects. Such inhibitors are widely distributed within the plant kingdom and can accumulate to high levels in seeds and storage organs. Other major advantages of protease inhibitors is their inactivation with cooking.

In cowpea (*Vigna unguiculata*), trypsin inhibitor (CpTi) level was shown to be responsible for its resistance to attack by the major storage pest of its seeds (i.e. bruchid beetle = *Callosobruchus maculatus*) A variety of insects have shown to be toxic to CpTi (Table 2.3). This CpTi gene was joined with a CaMV 35S promoter, and one marker gene with the transgene and was used to infect tobacco leaf discs. *Agrobacterium* was used for transformation and the transgenic tobacco plants express a high level of CpTi. The CpTi gene in transgenic plants is stably inherited and there is no serious 'yield penatly'. Thus like Bt toxin, CpTi can also be used as a protectant against insect attack in transgenic plants. However, extensive field trials are necessary before releasing these transgenic plants to the farmers.

Gene for other insecticidal secondary metabolites

The production of secondary metabolites by plants can also implicate the resistance to insect attack (Table 2.4). However, biosynthesis of each of these metabolites involves many steps, each controlled by a separate gene. Furthermore, these genes are tissue specific in expression. These features make the production of transgenic plants difficult in this case. Utilizing these facilities, transgenic plants with insect resistance due to secondary metobolites will be available in the near future.

Table 2.4: Examples of secondary compounds from legume seeds with demonstrated insecticidal properties.

Compounds	Bruchid (insect)	Legume
Non-protein antimetabolites		
Alkaloids		
(i) 2,5-dihydroxymethyl-3,4-dihydroxypyrrolidine (DMDP)	*Callosobruchus maculatus*	*Lonchocarpus* sp.
(ii) Castanospermine	*Callosobruchus maculatus*	*Castamospermum austrole*
Non-protein amino acids		
p-Aminophenylalanine	*Collosobruchus maculantus* *Zabrotes subfasciatus*	*Vigna sp.*

Compounds	Bruchid (insect)	Legume
Rotenoids	*Callosobruchus maculatus*	*Lonchocarpus salvadoreensis*
Saponins	*Callosobruchus chinensis*	*Phaseolus vulgaris*
	Callosobruchus chinensis	*Glycine max*
5. Polysaccharides		
Pectosans	*Callosobruchus chinensis*	*Phaseolus vulgaris*
Heteropolysaccharide	*Callosobruchus chinensis*	*Phaseolus vulgaris*
Polysaccharide (?)	*Acanthosclides obtectus*	*Phaseolus vulgaris*
	Callosobruches chinensis	*Vigna spp.*
Protein antimetabolites		
Lectins (phytohaemagglutinins)	*Callosobruchus maculatus*	*Phaseolus vulgaris*
α-Amylase inhibitors	*Zabrotes subfascitus*	*Phaseolus vulgaris*
	Callosobruchus chinensis	*Phaseolus vulgaris*
Protease inhibitors	*Callosobruchus maculatus*	*Vigna unguiculata*
Arcelin / LLP	*Zabrotes subfascitus*	*Phaseolus vulgaris*
	Zabrotes subfasciatus	*Phaseolus vulgaris*

ENGINEERED RESISTANCE AGAINST VIRAL INFECTION

Over seven hundred plant viruses are recognised causing various diseases and significant crop losses. Due to the lack of 'viricides', the practical control depends on methods that prevent or restrict virus infection. Various strategies have been applied for protecting viral infection; few are described here like:

Cross Protection

If susceptible strain of a crop is inoculated with a mild strain of a virus, the susceptible strain, as a result, develops resistance against more virulent strains. The phenomenon is known as **cross protection.** This method has been used to reduce yield losses in crops like tomatoes against **tomato moisac virus** (TMV), in potato against **potato spindle tuber viroid** and in citrus against **citrus tristeza virus**. In most cases of cross protection, the symptoms of infection are delayed, and even the replication of virus is suppressed, although eventually the severe strain may be able to overcome the protection.

The disadvantages of practicing cross protection includes

1. Possibility of mutation in inducing mild virus strain.
2. Possibility of synergism between inducing virus and another unrelated virus.
3. Possibility of unnecessary spread of mild virus causing threat for future yield losses.
4. Possibility of some yield losses due to mild strain. Transgenic plants have been produced in tobacco, tomato, and potato with single gene using a broad spectrum of plant viruses.

Coat protein mediated protection

Coat protein - mediated resistance to viruses has been one of the successes of plant genetic engineering. Coat protein gene from tobacco mosaic virus (TMV), a positive strand RNA virus, has been transferred to genome of tobacco plants. In the transgenic plant, expression of coat protein (CP), as well as, low and delayed infection was observed when inoculated with TMV. However, in such transgenic plants, coat protein gene should be constitutively expressed and may thus have effects on the nutritional value of plants. Subsequently Cp genes isolated from different plus strand RNA plant viruses have been cloned into plant expression vectors and transferred to various crop plants like tobacco, tomato and potato, using *Agrobacterium* mediated plant transformation system. Potatoes have been produced which have coat protein genes and are tolerant in field tests to both potato virus X (PVX) and potato virus Y (PVY). A target DNA coding for a component of TMV replicase enzyme, was also transferred to tobacco plants, conferring resistance to TMV.

Antisense RNA mediated protection

Antisense constructs also generate protection against RNA and DNA viruses. Transgenic tobacco plants expressing antisense transcript (antisense - coat protein gene) were protected from CMV infection.

Transgenic tobacco plants with Nucleocapsid (N) gene, CaMV 35S promoter and a leader sequence (to enhance expression) showed expression of N-gene. Such transgenic plants were resistant to tomato spotted wilt virus. (TSWV). TSWV is a negative strand RNA virus with RNA joined firmly to the nucleocapside protein.

Satellite RNA mediated protection

Satellite RNAs are associated with specific strains of some plant RNA viruses, though they are unrelated to viral genome. These extragenomic RNA species depend on a helper virus for replication. Some satellite RNAs are known to attenuate the symptoms of their helper virus. Thus satellite RNA might be used to protect crop plants against the effect of CMV infection. The nature of protection in transgenic plants by satellite RNA is different from coat-protein protection.

Defective Interfering mediated protection

Defective interfering (DI) particles are deletion mutants of genomic viral sequences, which depend on their parent virus for replication. Like satellite RNAs, DI particles can attenuate the disease symptoms of their helper virus by interfering with its replication. Artificial DI molecules of TMVY, TMV and BMV have been constructed and inserted in plant protoplast e.g., deletion mutants of BMV RNA 2 exhibited reduced replication of MBV RNA1 and RNA2 in barley protoplasts.

Replicase mediated protection

It has been discovered that transgenic tobacco expressing part of replicase gene of TMV strain was highly resistant to this and closely related strains.

Currently, atleast three approaches are available to broad spectrum resistance. The first is based on the replication and cell to cell movement of virus. Second approach relies on the transgenic expression of antivirus protein of non-plant origin including antibodies. The third approach is based on genetic manipulation of natural viral defence in plants.

ENGINEERED RESISTANCE AGAINST FUNGAL PATHOGEN

Genetic engineering has opened vistas with isolation of resistance genes and gene transformation methods to develop transgenic plants. Resistance to fungal pathogen is achieved by isolating and transferring fungal resistance genes to the desired susceptible crop. Compounds toxic to fungus can also be used to produce transgenic plants as done in case of tobacco. Antifungal proteins like **endochitinase** and **ß-1,3-endoglucanases** are thought to inhibit fungal growth. Transgenic tobacco plants with endochitinase CH5B gene with Ca MV 35S promoter exhibits reduced seedling mortality when grown in presence of *Rhizoctonia solani.*

Another method is using ribosome inactivating proteins (RIPs). These RIPs do not interfere with the functioning of self-ribosomes but exert inhibitory activity towards fungal ribosomes. When RIPs are mixed with endochitinase or ß-1,3-glucanases, the antifungal activity exhibits synergistic increase in *in vitro* assay. Transgenic tobacco with barley type RIP gene and potato Wun1 promoter showed an increased resistance to *R. solani.* Table 2.5 gives a brief account of transgenic crops showing resistance against bacterial and fungal pathogens.

Table 2.5: Some pathogens for which resistance has been transferred in some crop plants

Pathogen	Disease	Resistance gene	Source of gene	Transgenic crop
Pseudomonas syringae	Wild fire	Acetyl transferase gene		Tobacco
Alternaria longipes	Brownspot	Chitinase gene	*Serratia marcescens* (soil bacterium)	Tobacco
Rhizoctonia solani		Chitinase gene	Bean	Tobacco
Phytophthora infestans	Late blight	Osmotin gene	Potato	Potato

GENETIC ENGINEERING AND PLANT LIPIDS

Plant lipids are major components of the human diet and are present in plants as storage oil. Engineered vegetable oils have been tested as field trials. A high stereate phenotype has been engineered in rapeseed by antisense approach. Stearoyl-acyl carrier protein (Stearoyl - ACP) desaturase is a chloroplast enzyme which catalyzes the first desaturation step in seed oil biosynthesis, converting stearoyl-ACP to oleayl-ACP. This coding region of the gene for rapeseed desaturase was fused in an inverted position to the napin storage protein promoter and to the promoter of seed-specific acyl carrier protein gene, and then introduced into rapeseed. This reduced the activity of desaturase enzyme in engineered plant resulting in increased levels of stearate in seeds (upto 40% of the total fatty acids), with corresponding decrease in oleate. Due to tissue specific expression of antisense RNA, integrity of membrane lipids in leaf remained unaffected.

Length of fatty acids can be manipulated with thioesterase enzyme, which is responsible for the accumulation of short chain fatty acids in triglycerides. The gene has been cloned and expressed in developing seed of *Arabidopsis*, where it caused the accumulation of C12 storage lipids. It is useful for application in confectionery production.

The field of storage lipid modification via the engineering of the fatty acid biosynthesis enzymes (Table 2.6) promises to be a very fertile one that will contribute to production of healthier food and of chemical feed stocks.

Table 2.6: Modifications of lipid metabolism in transgenic plants

Modification achieved	Enzyme engineered	Source of gene
Lauric acid production	Acyl-ACP thioesterase	*California bay*
Increased Stearic Acid	Antisense of stearoyl ACP desaturase	*Brassica napus*
	Stearyl-CoA desaturase	Rat
Reduced saturated fatty acids	Stearyl-CoA desaturase	Yeast
	3-Ketoacyl-ACP synthase II	Castor
	Acyl-ACP thioesterase	Soybean
Reduced saturated fatty acids in phosphatidyl glycerol	Acyl-ACP glycerol-3-phosphate acyltransferase	Squash, *Arabidopsis*
Increased saturated fatty acid in phosphatidyl glycerol	Acyl-ACP glycerol-3-phosphate acyltransferase	*Escherichia coli*
Petroselinic acid production	Acyl-ACP desaturase	Coriander
Increased-linolenic acid	ω3 desaturase	Arabidopsis
Cylopropane fatty acid production	Cylopropane synthase	*Escherichia coli*
γ-Linolenic acid production	Linolenic Δ6 desaturase	Synechocytis
Increased oleic acid	ACP: protein a fusion	Spinach

GENETIC ENGINEERING AND STORAGE PROTEINS

Human nutrition requires a balanced source of amino acids and the amino acid balance of many plant products is unsatisfactory. Genetic engineering may be carried out in crops making following three approaches: (a) expressing a desirable and heterologous storage protein, (b) increasing the level of a desirable, but little expressed, endogenons protein and, (c) suppressing the expression of antinutritional proteins.

A heterologous species of sulphur-rich storage protein in Brazil nut containing 18% methionine and 8% cysteine has been expressed in tobacco and rapeseed. In transgenic tobacco, **phaseolin** gene has been expressed, leading to accumulation of methionine rich phaseolin.

Another approach for increasing amino acid synthesis is by engineering of a rate-limiting enzyme.. For example, in a tobacco mutant lysine overproduction was caused by a lysine-insensitive **dehydro-picolinate synthase.** Forty fold increase in free lysine has been obtained by expression of a bacterial dehydropicolinate synthase in plant chloroplasts.

Genomic clones encoding two subunits of soybean **β-conglycin** (7S protein) genes were transferred into petunia plants. The temporally regulated expression of these two genes in transgenic plants mimic their temporal regulation during soybean embryo development. The gene encoding for 42 kDa potato storage protein, **patatin** was introduced into tobacco. Stable synthesis of patatin protein was confirmed in the heterologons environment. Furthermore, transgenic tobacco possessing the 5' and 3' flanking sequences of the bean ß-phaseolin gene fused with 15 kDa *zein* gene could synthesize up to 1.6% of total seed protein as zein. Thus, zein is deposited and accumulates in vacuolar protein bodies of the tobacco embryo and endosperm.

GENETIC ENGINEERING FOR NITROGEN FIXATION

Transfer of nif-genes from *Klebsiella pneumoniae* to *E. coli*

Klebsiella pneumoniae is a non symbiotic nitrogen fixing bacterium. R.A. Dixon and J.R. Postgate (1972) transferred nif-genes of *K. pneumoniae* into *E. coli* cells by conjugation. The transgene in the new host produced functional nitrogenase system. The nif-genes of *K. pneumoniae* have been introduced into several other organisms, e.g. *K. aerogenes*, *Salmonella typhimurium*, *Erwinia herbicola*, *Serratia marcescens*. These organisms are genetically modified and are now able to fix atmospheric nitrogen.

Transfer of *nif*-genes of eukaryotic organisms

Nif-genes from Rhizobium have been transferred to *Agrobacterium tumefaciens*. The expression of transferred gene has been successful in cereal crops like wheat, paddy etc. through the Ti-plasmids of *A. tumefaciens*.

Transfer of nod-genes to increase host range

Cloned nod-genes of *Rhizobium leguminosarum* that nodulate pea have been transferred to *R. phaseoli* (this normally nodulates bean). The transformed *R. phaseoli* can now nodulate pea plants also.

Plants that produce plastic

Chris Somerville and coworkers have evolved a transgenic *Arabidopsis thaliana* to polyhydroxy butyrate (PHB). It is a natural and totally biodegradable plastic material. In future, plastic producing sugar beet and potatoes could also be produced. These scientists have isolated two genes from a bacterium *Alcligenes eutrophus* which produced PHB in nature and accumulates it as a coating. These two genes were inserted separately into two *A. thaliana* plants with the help of *Agrobacterium tumefaciens*.

MISCELLANEOUS APPLICATIONS

The knowledge of genetic engineering has been exploited in industries to deliver products and processes of economic and social value. Products of biotechnology include organisms and substances made by it that are consumed directly as such or serve as components for other industries. Processes of social value include technologies aimed at social hygiene and well-being, such as purification of drinking water, pollution control and waste disposal. The actinomycetes or soil microorganisms that produce the plethora of miracle drugs - the **antibiotics** have also been added to the list. The new techniques permit harnessing of single cells to produce non-native gene products from exogenously supplied genes. Industrial use of single cells and enzymes for the production of stated products may require special tailoring of the cells for adaptation to mass-scale production processes.

The pharmaceutical and chemical industries were the first to exploit the gene engineering technology. A futuristic application is envisaged in the field named **bioelectronics** or **bionics**. The potential of biological material to serve as extra-sensitive conductors, sensors, and support matrices for semiconductor molecules is already being explored. The drugs and pharmaceuticals may be synthesized in heterologous cells or in animal cell lines that produce them. For instance, human chorionic gonadotropin (HCG), erythropoietin and calcitonin synthesizing cell lines have been reported. Of these, only HCG is successfully used for commercial production. The drugs and pharmaceuticals that are candidates for the genetic engineering are:

1. hormones
2. growth factors
3. blood-derived factors
4. enzymes
5. antibiotics
6. immunoproteins
7. immunodiagnostics
8. DNA probes for molecular diagnosis

3

Gene Isolation and Identification

Recombinant DNA technology deals with isolation and manipulation of genes. Before nucleic acids can be cut or otherwise manipulated, they must be either isolated and purified or artificially synthesized. The isolated genes are also sometimes sequenced for a better understanding of its structure that needs to be manipulated. In this chapter we will discuss the methods available for isolation, sequencing and synthesis of genes.

ISOLATION OF TOTAL CELLULAR DNA

DNA is very easily damaged by shear forces; even rapid stirring of solution can break even with high DNA molecular weight into much shorter fragments. Consequently, DNA is recovered from cells by gentlest possible method of rupture, in the presence of EDTA to chalate the Mg^{2+} ions needed for DNase activity. Cell walls, if present, are digested enzymatically (e.g. lysozyme treatment of bacteria) and the cell membrane is solubilised using detergent. Cell disruption is performed at 4°C using autocloved glassware and solutions to destroy DNase activity.

After release of nucleic acids from the cells, RNA can be removed by treatment with ribonuclease (RNase). The other major contaminant, protein, is removed by shaking the solution gently with water-saturated phenol, or with phenol/chloroform mixture, either of which will denature protein but not nucleic acids. Centrifugation of the emulsion formed by this mixture produces a lower, organic phase, separated from the upper aqueous phase by an interface of denatured proteins. The aqueous solution is recovered and deproteinised repeatedly, until no matter is seen at interface. Finally, the deproteinised DNA preparation is mixed with ethanol and allowed to precipitate out in freezer. After centrifugation, the DNA pellet is redissolved in a buffer containing EDTA for protection against DNases and this solution can be stored at 4°C for at least a month. DNA solutions can be stored frozen, but repeated freezing and thawing tends to damage long molecules by shearing; so preparation in frequent use are normally stored at 4°C.

This procedure is suitable for total cellular DNA. If the DNA from specific organelle or viral particle is needed, it is best to isolate the organelle or virus before extracting its DNA. Now we will discuss several basic techniques related with gene isolation and identification.

NUCLEIC ACID HYBRIDIZATION

If a double-stranded DNA molecule is exposed to high temperature, or to very alkaline conditions, then the two strands will come apart. The molecule is said to have become denatured. The temperature at which denaturation occurs is termed as melting temperature or T_m. If the denatured DNA is returned to a temperature below its T_m, or to neutral pH when alkali was used to denature it, each strand will, after a time, find its complementary strand. The two strands will 'zipper' back together to re-form a double-stranded DNA molecule. This ability of complementary sequences to anneal, or hybridize, to one another is called nucleic acid hybridization. This techniques helps in determining the gene structure and in identifying molecules which contain same sequences of nucleotides. In a complex mixture of nucleic acid molecules, nucleic acid hybridization technique helps in separation of complementary sequence.

When employed analytically, hybridization is normally performed using one labelled sequence, termed the probe, and an unlabelled sequence called the target. **Probe** is a short synthetic oligo deoxyribonucleotide which is complementary to target DNA sequences. Target DNA may be obtained from total cellular DNA preparations. The probe is labelled by incorporation of either radioactively labelled nucleotides or with some chemicals (described later). The probe is the known, pure species in the hybridization and the target is the unknown species to be identified.

The target will most often form part of a mixture of unrelated nucleic acid sequences. The probe is usually added in considerable excess over the target, and its concentration determines the rate of reaction. Hybridization reaction time is extended to ensure complete reaction. The target and probe can be hybridized to each other in solution, or alternatively the target can be immobilized on an inert support such as nitrocellulose (Fig 3.1). When the target is immobilized the rate of annealing is greatly reduced.

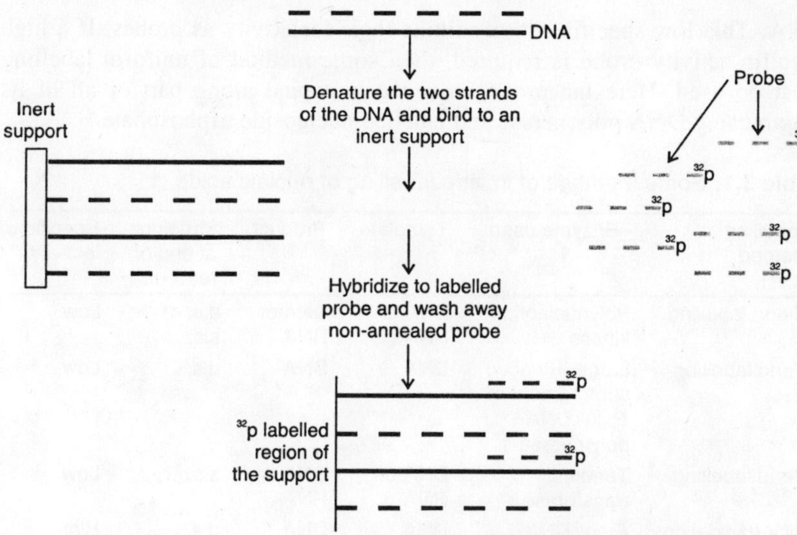

Fig. 3.1: The principle of hybridization using an inert support. The target DNA is most often localized to one part of the support and, in most cases, the aim of the hybridization experiment is to identify this region. The probe is prelabelled by some method, most often by the incorporation of radioactivity. After hybridization, the annealed probe is detected by autoradiography.

The rate of hybridization depends upon the DNA concentration, the concentration of salt in the solution and the temperature. Hybridization reactions are normally performed at 10-20°C below the T_m of the duplex. Under any given conditions of ionic strength and temperature, the T_m of a duplex also depends on the base composition of the annealed partners.

Hybridization involving RNA are performed in the presence of formamide as RNA cannot withstand higher temperatures. Formamide disrupts hydrogen bonding and thus reduces the effective T_m of the duplex, allowing annealing to occur at a lower temperature (e.g. 37°C).

METHODS FOR LABELLING NUCLEIC ACIDS

For identification of hybridized nucleic acid duplexes, labelling of probe or target is necessary. As probe is short synthetic oligonucleotide it can be easily labelled. A ^{32}P isotopic label can be added to the 5' hydroxyl group using polynucleotide kinase to a probe (Table 3.1). Alternatively, labelled nucleotides can be incorporated at the 3' end using terminal transferase. Because label is added at only one place within the oligonucleotide, such end-labelled probes contain relatively little radioactivity per unit weight of

DNA. This low specific activity limits their sensitivity as probes. If a high specific activity probe is required, then some method of uniform labelling must be used. Here the probe sequence is copied along part or all of its length using DNA polymerase and labelled nucleoside triphosphate.

Table 3.1: Some methods of in vitro labelling of nucleic acids.

Labelling method	Enzyme used	Template	Product	Structure at end of reaction*	Specific activity
5' end labelling	Polynucleotide kinase	DNA or RNA	DNA or RNA	d.s. or s.s.	Low
3'end labelling	Large (Klenow) subfragment of *E.coli* DNA polymerase-1	DNA	DNA	d.s.	Low
3'end labelling	Terminal transferase	DNA or RNA	DNA or RNA	s.s.	Low
Nick translation	*E.coli* DNA polymerase-1	DNA	DNA	d.s.	High
Random priming	Klenow subfragment of DNA polymerase-1	DNA or RNA	DNA	d.s.	High
In-vitro transcription	Bacteriophage (e.g. T7) RNA polymerase	DNA	RNA	s.s.	High

*d.s. - double stranded; s.s. - single stranded.

METHODS OF LABELLING NUCLEIC ACID AND PROBES

There are five basic methods for labelling nucleic acids. These are: nick translation, primer extension, methods based on RNA polymerase, end labelling methods and direct labelling methods.

Nick translation

This is done by making single-strand cuts ("nicks") in the double-stranded DNA molecule by brief exposure to a dilute solution of an endonuclease (usually deoxyribonuclease I of *E.coli*). DNA polymerase I is then used in the presence of at least one radioactive precursor (e.g., Â-^{32}P-dATP) to "translate" the nick along the molecule in the 5' to 3' direction. This occurs by the concerted action of the 5'→3' polymerase and the 5'→3' exonuclease activities built into DNA polymerase I. The 5'→3' polymerase adds nucleotides to the 3'-OH terminus at each nick, while the 5'-phosphate

terminus at the nick. The net result is that a nonradioactive strand of DNA is replaced by a radioactive strand (Fig 5.2). The DNA is then denatured and used as a radioactive probe in hybridization experiments (Southern blots, northern blots, etc.).

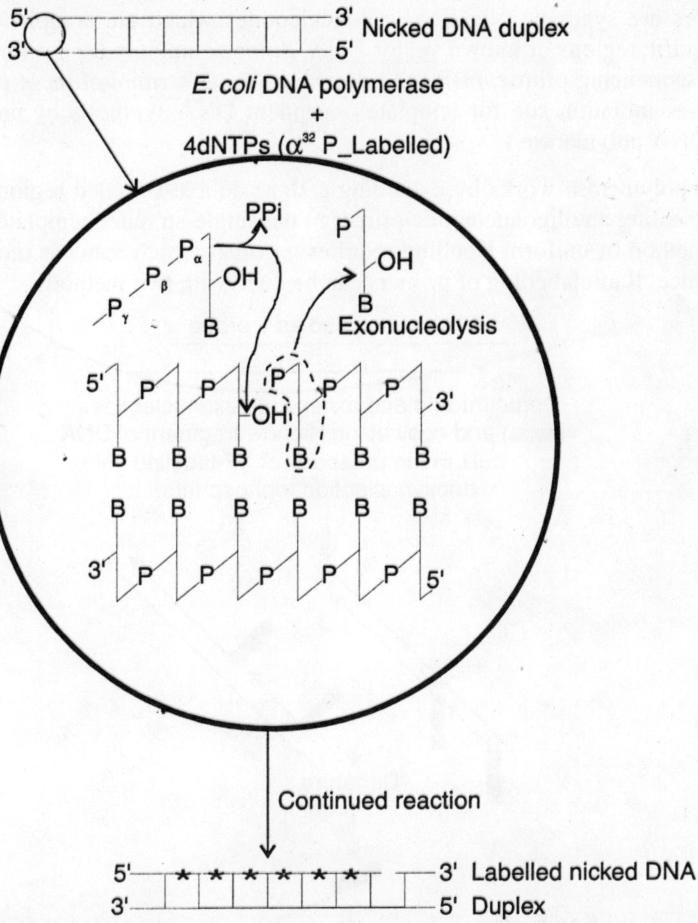

Fig. 3.2: ^{32}P-labelling of duplex DNA by nick translation. Asterisks indicate radiolabelled phosphate groups.

Nick translation can be used with a variety of labels to generate probes suitable for most hybridization applications. It is also appropriate for the generation of biotinylated probes (described later).

A major advantage of nick translation over other uniform labelling methods is its flexibility with respect to probe size, specific activity and concentration. It is particularly suitable for the production of large quantities of probes for use in multiple hybridization and/or where a high probe concentration is required.

Primer Extension Method

Primers are synthetic oligodeoxyribonucleotides which are complementary to specific regions of known vector DNA. Some commonly used primers are M13 sequencing primer, pBR322 primer etc. The 3' termini of these primers serve as initiation site for template dependent DNA synthesis by enzymes like DNA polymerase I.

DNA polymerase works by extending a short double-stranded region made by annealing an oligonucleotide primer to the single-stranded template. Thus this method of uniform labelling requires a primer which matches the probe sequence. Radiolabelling of primers can be done with two methods.

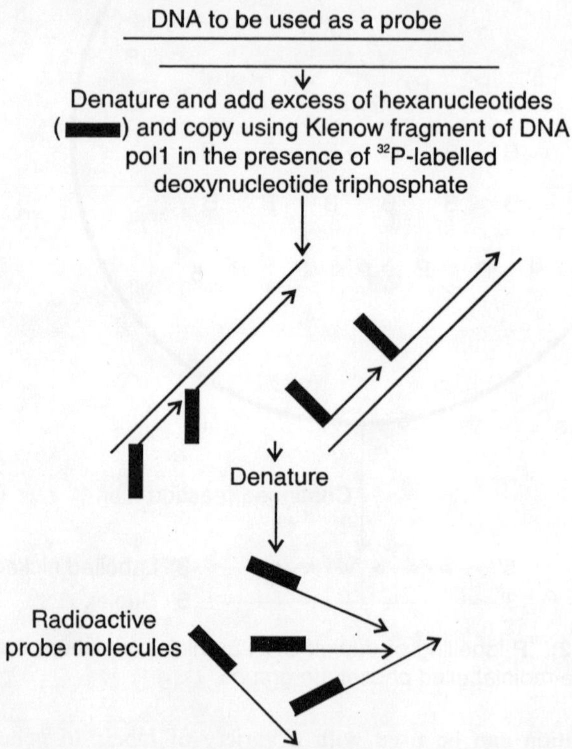

Fig. 3.3: The principle of random primed (oligo-) labelling. The DNA to be used as a probe is denatured by heating and mixed with hexanucleotides of random sequence which then act as primers for DNA synthesis.

If the probe sequence is not known then random oligonucleotide labelling can be used (Fig. 3.3). It is often in the case when natural cellular DNA is used. These primers are made by adding a mixture of all four bases at each step in the chemical synthesis reaction. The DNA is denatured and the two complementary strands are copied in the presence of labelled primers as well as, nucleoside triphosphates. The polymerase used is Klenow fragment derived from DNA polymerase-I of *E. coli*. Chance homology ensures that these primers anneal to the separated DNA strands at many points along their length, thus providing a base for polymerase to initiate DNA synthesis. This is only one of several uniform labelling methods.

The second method uses a unique primer to restrict labelling to a particular sequence of interest. In the primer extension method it is essential to use a polymerase lacking a 5'→3' exonuclease activity otherwise degradation of the primer will occur. Both the Klenow fragment of *E. coli* DNA polymerase I, which lacks the 5'→3' exonuclease activity and reverse transcriptase have been used successfully.

An advantage of random priming method is that is may be used to label relatively impure DNA. In primer extension method, there is not synthesis of DNA and labeled nucleotides are incorporated in concentrations. It is an ideal method for situations where high specific activity and low probe concentration are frequently employed.

Methods based on RNA polymerases

RNA polymerases catalyzes the synthesis of RNA from nucleoside triphosphates using a DNA template. Thus they can incorporate labelled ribonucleotides into RNA during transcription if such labelled nucleotides are provided to it. If a specific site of a vector or DNA is transcribed in such way, RNA probes (or transcripts) of defined length and sequence can be obtained.

Initially RNA probes were obtained from RNA polymerase from *E. coli*. This enzyme when used *in vitro* exhibited very little template and promoter specificity and therefore produces transcripts which have been initiated and terminated at random. Today, RNA polymerases from a number of phages have been used, which possess a high degree of specificity. Thus transcription can be limited to sequences cloned downstream of an appropriate promoter.

RNA probe's specific activity is determined solely by the specific activity of labelled nucleotide and moreover such probes are need not to be denatured before hybridization. These probes are more sensitive than nick translated probes because of avoidance of probe reannealing during hybridization (as RNA-RNA hybrids are more stable than RNA-DNA hybrids). A further advantage of RNA probes is that non specifically bound probe can be removed with RNase A. RNase A is very specific for single stranded RNA.

The only difficulty with RNA polymerase-based-labelling is that a specific promoter is read by RNA polymerases before transcripting the downstream. This promoter has to be subcloned and inserted into the vector. However, once an insert has been cloned into such a vector, DNA probes may still be prepared using nick translation or random primer labelling.

END-LABELLING OF NUCLEIC ACIDS

A wide variety of techniques is available for introducing label at either the 3' or 5' ends of linear DNA or RNA. Usually only a single label is introduced at the terminus. Nucleic acid can be 5' end labelled using T4 polynucleotide kinase. The enzyme catalyzes the transfer of the γ-phosphate of ribonucleoside 5'-triphosphate donor to the 5'hydroxyl group of a poly- or oligo-nucleotide or even to nucleoside-3-triphosphate. Radiolabelled phosphate group is donated by [γ^{32}-P] ATP to DNA or RNA containing a 5'-hydroxyl terminus. This is termed as forward reaction.

If 5'-phosphate group is present in DNA or RNA, then it is removed with alkaline phosphatase. This reaction is driven by excess ADP which causes the enzyme to transfer the terminal 5'-phosphate from DNA to ADP (Fig. 3.4). This is known as exchange reaction. The DNA is rephosphorylated by transfer of labeled γ-phosphate from [γ^{32}-P] ATP.

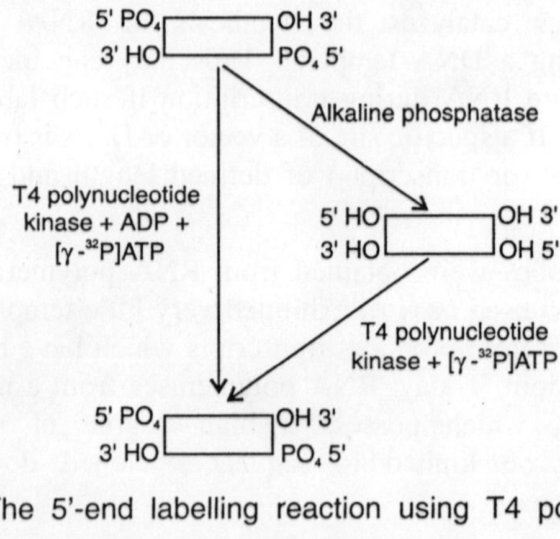

Fig. 3.4: The 5'-end labelling reaction using T4 polynucleotide kinase.

The major advantages of 5'-end labelling are:

1. Both DNA and RNA can be labelled.

2. Oligonucleotide can be conveniently labelled e.g. for detection of mutants following site-directed mutagenesis.
3. Location of labelled group is known.
4. Very small fragments can be labelled.
5. Restriction digest fragment can be labelled.

DNA fragments also can be labelled at their 3'end by using terminal deoxynucleotidyl transferase, however, this method is not of wide use.

CHOICE OF LABEL

There are two technical parameters, resolution and sensitivity, which determines the success of probe application. High degree of resolution is required to know the relative position of a nucleic acid fragment. High sensitivity is necessary because sequence of interest may be present at low abundance.

Other factors are probe stability, safety and ease of use. Broadly labels can be categorised into radioactive and non radioactive types.

Radioactive labels

These labels have wider applications as they can be easily detected with autoradiography. Their detection gives two important information, firstly about occurrence of hybridization between probes and target DNA and secondly about their position. Radioactive methods using ^{32}P are easily detectable and are used often.

Non radioactive labels

A number of non radioactive labels for probes are available but biotin is widely used. Brief account on biotin and other non radioactive table is given below:

(a) *Biotin labelled probes*. Biotinylated probes are prepared through a nick-translation reaction by replacing nucleotides with biotinylated derivatives. After hybridization and washing, detection of hybrids is done by adding avidine and going through a series of cytochemical reactions which finally give a blue colour whose intensity is proportional to the amount of biotin in the hybrid. There are several advantages of using biotinylated probes. The major advantages of using biotinylated probes are: (a) assays employ non- toxic materials, with longer half-life (b) can be prepared in advance in bulk and stored at -20°C for repeated uses (c) detection of hybrids is much faster than by radioactive probes. A limitation of this technology is that biotinylated sites are limited, thus decreasing the intensity of signal. It has been solved by adding long 'tails' of biotinylated nucleotides to the probes through enzymatic methods. Another disadvantage of biotin labelled probes is that the cytochemical visualization reactions lead to precipitation of insoluble material which can not be removed and therefore, the filter can not be reused.

(b) *Digoxigenin labelled probes*. Another method is to label DNA with digoxigen-labeled deoxyuridive triphosphate (dig-ÂUTP). After hybridization to the target DNA the hybrids are detected by enzyme-linked immunoassay using an antibody conjugate e.g. anti digoxigenin alkaline phosphatase. Enzyme reporter molecules with enzymes can also be used as label.

(c) *Alternatives to biotin and digoxigenin labelling*. Other ligands (e.g. hapten determinants, 2, 4-dinitrophenol, arsenate derivatives, etc.) are attached to specific proteins that can be tagged with enzymes or fluorescent molecules. It is possible to monitor many probes simultaneously by using several different ligands since each ligand would yield a different signal. A chemiluminescent (light emitting) probe system has also been developed in which two different probes complementary to a continuous segment of DNA hybridize to adjacent of a gene. The first label is a chemiluminescent complex which emits light at a specific wavelength. This emission excites the label molecule on the second probe to emit light at a different wavelength, which can be detected using a photomultiplier device. The processes called non-radioactive energy transfer, and can occur only if the two probes hybridize correctly and the two labels are close to each other.

The major disadvantage of biotin, and of other nonradioactive labels, is low sensitivity in comparison to radioactive labels.

MAPPING GENES ON SPECIFIC CHROMOSOMES

One of the most important applications of nucleic acid hybridization is construction of genetic maps. This helps in determining the chromosomal location of individual genes. There are three commonly used methods that differ in the degree of resolution they afford viz somatic cell hybrids, *In situ*, hybridization and genetic linkage mapping.

Somatic Cell Hybrids

Somatic cell hybrids are prepared by fusing together human and rodent cells using reagents, such as polyethylene glycol or Sendai virus. The reagents help in membrane fusion. The resulting hybrid cells are then grown in culture medium containing selective agents, which allow only the hybrid cells to grow. In hybrids produced using human and rodent cells the rodent chromosomes are retained, while the human chromosomes are lost. This random chromosome loss occurs during continued culture and shows no apparent selectivity for individual human chromosomes. Thus it is possible to produce a panel of hybrid cells containing different combinations of human chromosomes. Genomic DNA is prepared from panels of such hybrids and used in hybridization analysis. The human probe to be mapped is hybridized with such hybrids. By determining which of the panel of hybrids shows a hybridization signal, and correlating this information with the karyotype of the cells, it is possible to deduce which chromosome carries the test gene.

In-situ Hybridization

In this technique a labelled DNA probe is hybridized to metaphase chromosome. The position of hybridization is visualized at the light microscope level. Metaphase chromosomes are liberated from cells and bound to the surface of a cover slip. The DNA within the chromosomes is denatured in such a way that the individual chromosomes remain visually identifiable. The probe is labelled, either by incorporation of a ^{3}H-nucleotide or non-isotopically. The chromosomes are incubated with the labelled probe for many hours and, after hybridization, the excess unbound probe is washed away and the annealed probe detected.

If radiolabelled, probe used then the slides are autoradiographed. Deposited silver grain after autoradiography will lie in the approximate area at which the probe bounds to the chromosome. The accuracy of mapping by this technique is dependent on the size of the grains which, in the case of small chromosomes, can cover a significant portion of the total chromosome length.

If the probe is labelled non-isotopically then it is possible to determine the precise position of hybridization with much greater accuracy. The signal is detected using a fluorescent molecule thus is called fluorescent *in-situ* hybridization (FISH) (Fig. 3.5). It is more sensitive than radiolabelling and can be used to give a fairly precise position of hybridization.

Thus FISH is particularly useful for determining the respective locations of two markers on the same chromosome. The two probes are labelled with different immunologically labelled adducts and a single hybridization reaction is performed. The two probes are then detected using fluorochromes which emit at different wave lengths when excited by UV light, such as rhodamine (red fluorescence) and fluorescein (green fluorescence). This technique has been used for probes which are separated by few hundred kilobases of DNA on same chromosome. The FISH technique can also be applied to interphase nuclei, which allows accurate ordering of probes along a chromosome when they are separated by no more than 20-30 kilobases

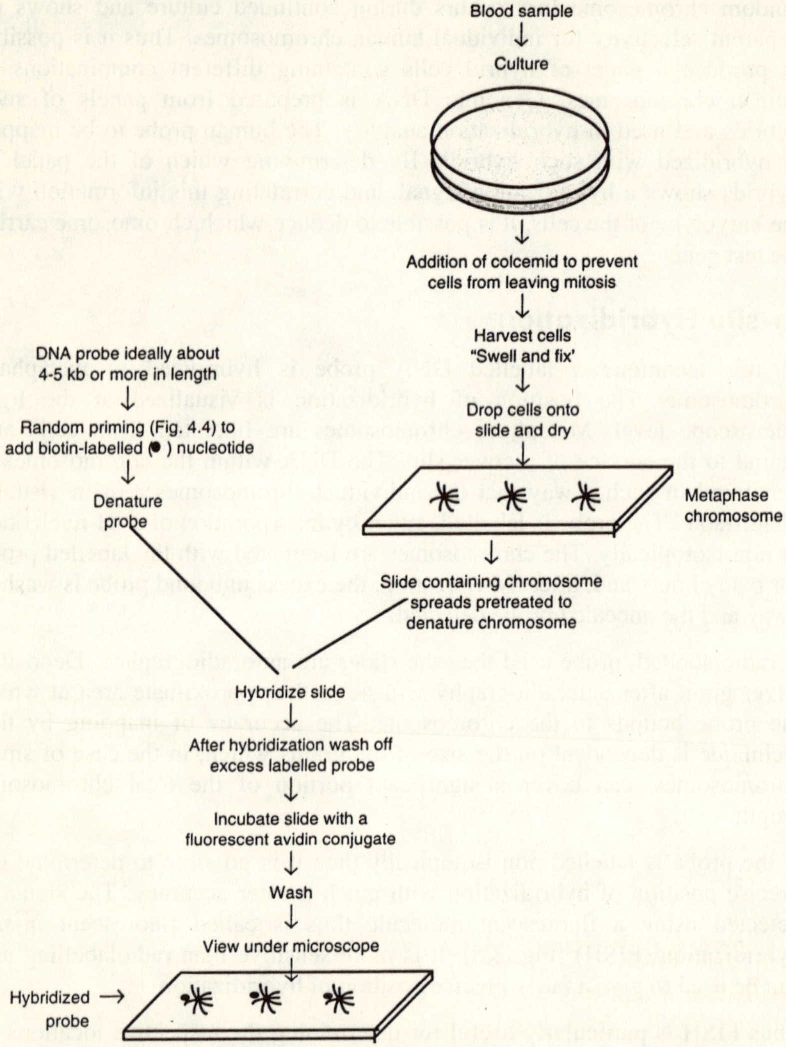

Fig. 3.5: FISH is used where gene mapping is done with in-situ hybridization. It uses a non-radioactive labelling system involving biotin attached. When viewed under special filters the specific sites of DNA probe can be seen on a chromosome.

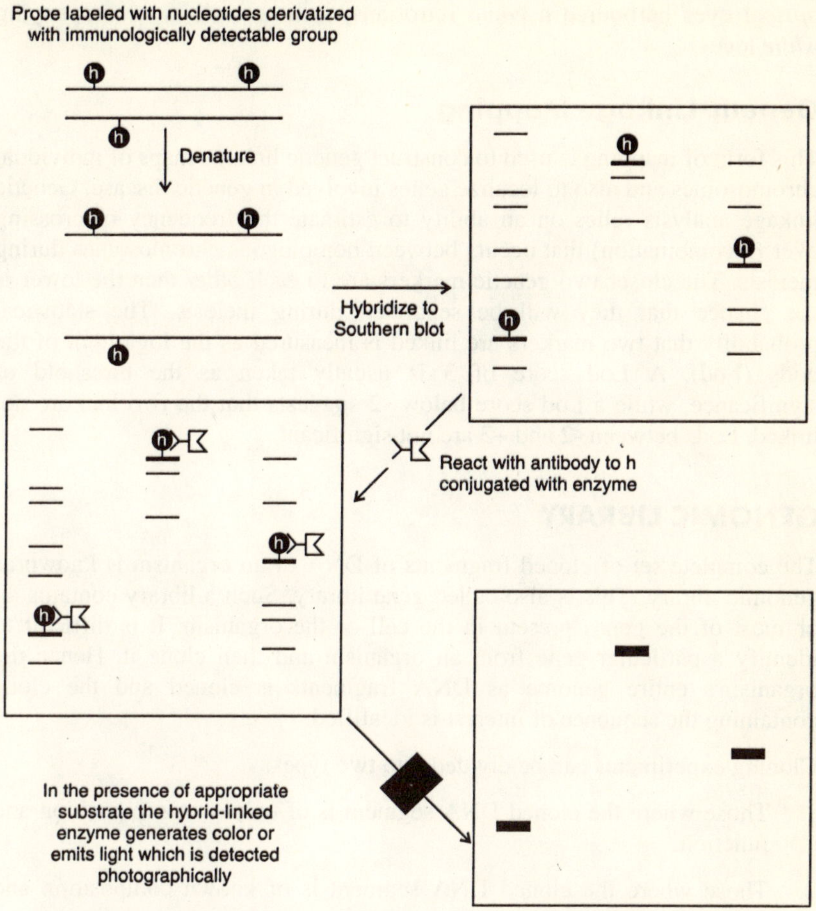

Fig. 3.6: The principle of non-radioactive detection methods for in-situ hybridization.

Transposon Tagging

Isolating a eukaryotic gene by transposon tagging involves inserting into target DNA, fractionating fragments of the latter on southern blots (described later in this chapter) and hybridizing the transposon with a probe. Cleaving out the transposon, together with flanking DNA, reveals the adjoining sequences of a gene. The probe are made for these adjoining sequences to find the fragment having gene of interest.

Presence of transposon is confirmed by altered phenotype. This method was first employed to isolate *white* locus in *Drosophila*. The mutants with *white-*

apricot eyes harboured a *copia* retrotransposon as well as accompanying *white* locus.

Genetic Linkage Mapping

This form of mapping is used to construct genetic linkage maps of individual chromosomes and also to localize genes involved in genetic disease. Genetic linkage analysis relies on an ability to estimate the frequency of crossing over (recombination) that occurs between homologous chromosomes during meiosis. The closer two genetic markers are to each other then the lower is the chance that they will be separated during meiosis. The statistical probability that two markers are linked is measured as the logarithm of the odds (Lod). A Lod score of 3 is usually taken as the threshold of significance, while a Lod score below -2 suggests that the two loci are not linked. Lods between -2 and +3 are not significant.

GENOMIC LIBRARY

The complete set of cloned fragments of DNA of an organism is known as genomic library. This is also called gene library. Such a library contains all or most of the genes present in the cell or the organism. It is difficult to identify a particular gene from an organism and then clone it. Hence the organism's entire genome as DNA fragments is cloned and the clone containing the sequence of interest is identified.

Cloning experiments can be divided into two types:

1. Those where the cloned DNA segment is of unknown composition and function.
2. Those where the cloned DNA segment is of known composition and function.

In the first approach the entire cellular DNA has to be fragmented into clonable size by either digestion with restriction endonucleases or, mechanical sheering.

These experiments with randomly cloned fragments are known as 'shot gun cloning experiments'.

The chromosomal DNA of the organism of interest is isolated. It is then treated with a known restriction enzyme to obtain DNA fragments of clonable size. The fragments are cloned using appropriate cloning vectors in any organisms like bacteria, virus or yeast. For identification, a marker gene is inserted while multiplying in vector. Hybridization of such genes can be carried out by two major techniques viz. colony hybridization and plaque hybridization depending on the type of vector used.

Colony hybridization

Once a genomic library or cDNA library is available, gene sequence can be isolated by colony hybridization techniques. It involves replica plating the colonies formed by transformed cell into nitrocellulose filter and their *in situ* hybridization with a radioactive probe. This method was first developed by Hogness and Grunestein (1975). Procedure is described in following steps:

1. The cloned yeast colonies, bacterial colonies or phage plaques to be tested are transferred from the culture plate on to a nitrocellulose filter paper by replica plating (Fig. 3.7).
2. The filter with the colony replicas is treated with NaOH to lyses the cells/phages and to denature DNA.
3. The filter is then baked to fix the DNA.
4. It is then treated with a radioactive ^{32}P labelled single stranded DNA probe that is complementary to the ssDNA of interest.
5. The filter is washed to remove the unbound excess probe, dried and then autoradiographed.
6. The DNA fragments hybridized with the radioactive probe can be detected on the autoradiograph film.
7. This indicates that the cells in the corresponding colonies contain the desired gene. The colony is taken out from the master culture plate for mass culture.

Plaque hybridization

This technique is applicable where a chimeric phage particles is carrying the gene of interest. In such a situation, a bacterial lawn is infected with a mixture of chimeric phage particles (i.e. the library) and a large number of plaques develop overnight. These plaques can be treated just like the colonies in colony hybridization to identify and isolate the chimeric phage particle carrying the gene of interest. Two new techniques have shown their potential in gene isolation and characterization. They are chromosome walking and chromosome jumping. Let us have a small account on both of these techniques.

Chromosome walking

Identification of fragments with overlapping sequences may be a key to the reconstruction or characterization of large chromosome regions. This is achieved by the technique popurlarly described as chromosome walking.

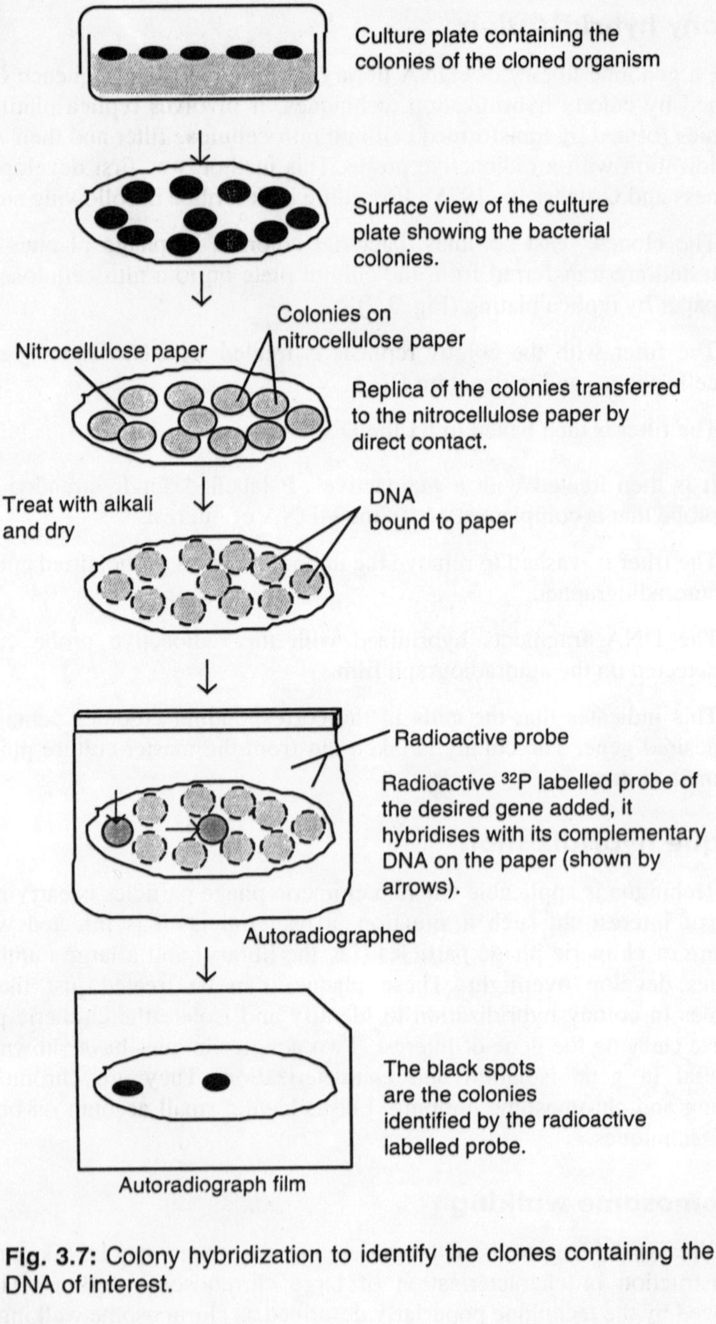

Fig. 3.7: Colony hybridization to identify the clones containing the DNA of interest.

The technique of chromosome walking involves the following steps:

1. From the genomic library a clone of interest (identified by a probe is selected.
2. One end of the clone is subcloned.
3. The subcloned fragment of the selected clone may be hybridized with other clones in the library and a second clone hybridizing with the subclone of the first clone is identified due to presence of overlapping region.
4. The end of the second is then subcloned and used for hybridization with other clones. This helps in identifying a third clone having overlapping region with the subcloned end of the second clone.
5. Identified third clone is also subcloned and hybridized with clones in the same manner and the procedure may be continued.
6. Restriction map of each selected clone is prepared and compared to know the region of overlapping, so that identification of few overlapping restriction sites resembles the walking along the chromosome or along a long chromosome segment.

Regions of chromosomes approaching 1000 kb have been mapped following the above technique. Restriction maps of entire chromosomes can be prepared in this manner following the technique of chromosome walking.

Chromosome jumping

In order to bring the molecular marker close to the gene of interest, *chromosome jumping* approach has been utilized. This technique can be described in the following steps:

1. The distance of 'jumps' or 'hopsize' (e.g. 100 kb or 200 kb) is determined.
2. Genomic DNA molecules in the range of selected size are selected through pulsed-field gel electrophoresis.
3. For circularization of DNA segments, ligation between two ends of each long linear DNA molecule is allowed using T4 ligase.
4. These circular DNA circles are.digested with *Eco*RI.
5. Phage vector Ch3A lac, is also cut with *Eco*RI and used for cloning small DNA fragments.

6. Such cloned DNA fragments represent the jumping library, which can be plated on a bacterial host and screened through the technique of plaque hybridization.

The technique of chromosome jumping helps in narrowing the gap between the gene and available molecular markers. After several cycles of chromosome jumping followed by cloning the regions broximal to the gene, it will be possible to approach very close to the desired gene capable approached and cloned. The gene for **cystic fibrosis** has recently been isolated using this technique.

CLEAVING DNA WITH RESTRICTION ENZYMES

After locating gene of interest in colony or on chromosome, the DNA is extracted and cleaved with enzymes. Most restriction enzymes cleave DNA molecules in a site specific manner. The genomic DNA from an individual organism is digested with either a single or more restriction enzymes. This results in fragmentation of DNA molecules with variation in length. These fragments are known as restriction fragment length polymorphisms or RFLPs. Since restriction enzymes cut DNA in a sequence specific manner, every homologous DNA molecule will be cut at exactly the same sites. This allows to isolate large quantities of specific DNA fragments for subcloning and isolating the DNA sequence of interest.

The sizes of the restriction fragments can be determined by polyacrylamide or agrose gel electrophoresis. It also helps in separation of various RFLPs. Here few electrophoritic and blotting techniques are described.

Gel eletrophoresis

Nucleic acid molecules can be distinguished by differences in their chain length, and study of gene organization and expression includes nucleic acid hybridization with a chain length measurement.

DNA and RNA have a net negative charge. At neutral pH therefore, these molecules migrate towards the positive terminal when placed in an electric field. This process of electrophoresis is performed in an agarose or polyacrylamide gel. Nucleic acids are loaded into slots in the gel and allowed to migrate towards the positive terminal. The pores in the gel act to sieve the molecules. So that the mobility of a particular nucleic acid species depends on its length. All the molecules of a particular size move at approximately the same rate through the gel, forming a band which gradually increases in width during electrophoresis because of diffusion. The size of the molecules can be estimated by running DNA molecules of known size in a parallel track on the gel and determining their migration position relative to the species of unknown size (Fig. 3.8).

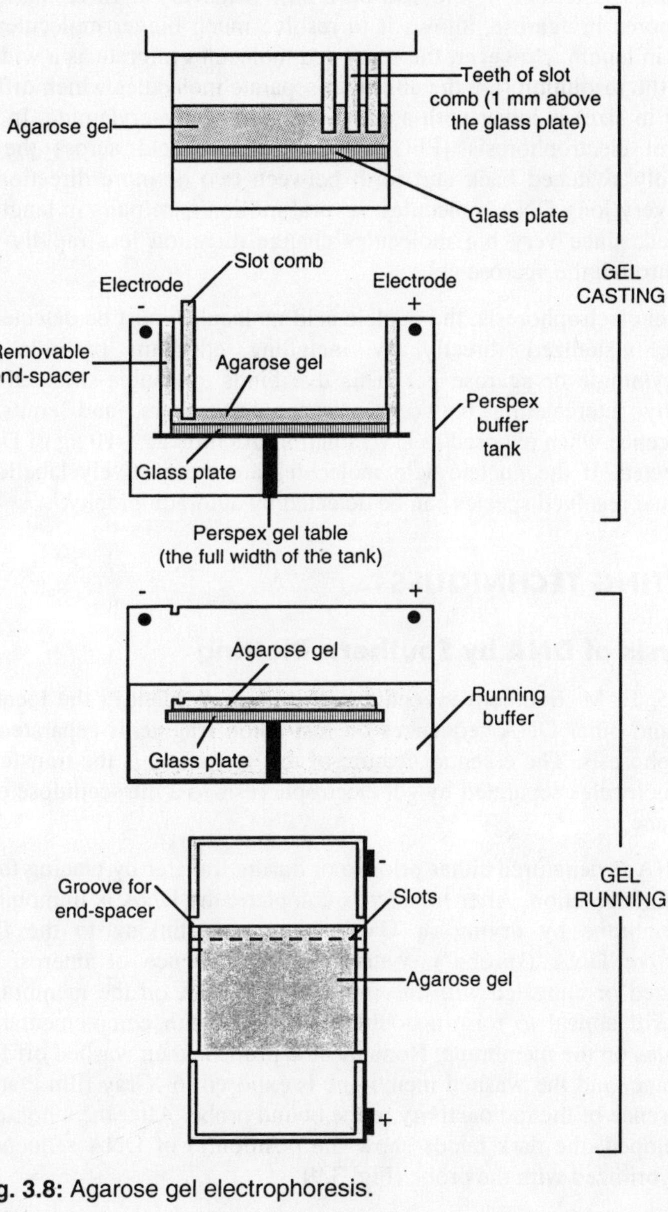

Fig. 3.8: Agarose gel electrophoresis.

Polyacrylamide has a smaller average pore size than agarose and so is effective at separating molecules of 10-1000 nucleotides in length (1000 nucleotides is termed 1 kilobase or 1 kb), with very high resolution. The larger pores in agarose, allows it to resolve much bigger molecules, up to 100 kb in length. However, the separated molecules migrate as a wider band so that the resolution (i.e. the ability to separate molecules which differ only slightly in size) is lower with agarose than with polyacrylamide. In pulsed-field gel electrophoresis (PFGE) the electrical field across the gel is repeatedly switched back and forth between two or more directions. This allows very long DNA molecules, several million base pairs in length, to be separated, since very big molecules change direction less rapidly as they move through the agarose gel.

After gel electrophoresis, the nucleic acid molecules must be detected. They can be visualized directly by including ethidium bromide in the polyacrylamide or agarose gel. This dye binds to double-stranded nucleic acids by intercalating between adjacent base pairs, and emits a red fluorescence when exposed to UV radiation. As little as 5-10 ng of DNA can be detected. If the nucleic acid molecules are radioactively labelled, then individual resolved species can be detected by autoradiography.

BLOTTING TECHNIQUES

Analysis of DNA by Southern Blotting

In 1975, E. M. Southern invented a procedure to identify the locations of genes and other DNA sequences on restriction fragments separated by gel electrophoresis. The essential feature of this technique is the transfer of the DNA molecules separated by gel electrophoresis to a nitrocellulose or nylon membrane.

The DNA is denatured either prior to or during transfer by placing the gel in an alkaline solution. After transfer is complete, the DNA is immobilized on the membrane by drying or UV-induced cross-linking to the filter. A radioactive DNA ("probe") containing the sequence of interest is then hybridized or annealed with the immobilized DNA on the membrane. The probe will anneal to form a double helix only with complementary DNA molecules on the membrane. Nonannealed probe is then washed off from the membrane, and the washed membrane is exposed to X-ray film that detects the presence of the radioactivity in the bound probe. After the autoradiogram is developed, the dark bands show the position(s) of DNA sequences that have hybridized with the probe (Fig. 3.9).

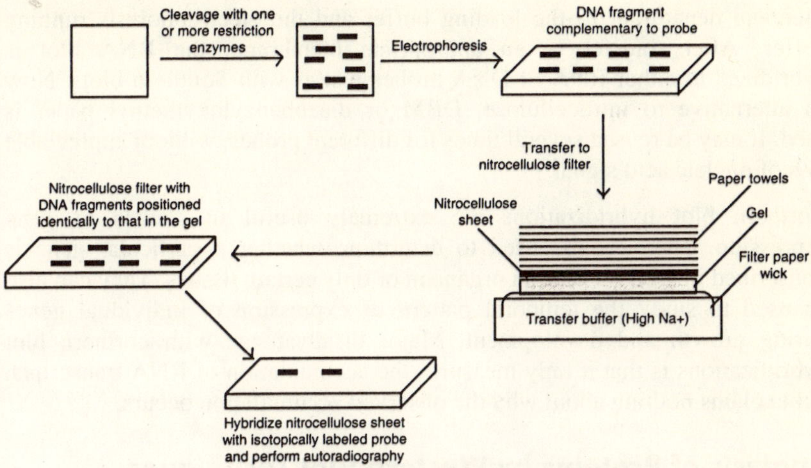

Fig. 3.9: Southern blotting method of analyzing DNA segment that share homology with a nucleic acid probe.

The ability to transfer DNA restriction fragments or other DNA molecules that have been separated by gel electrophoresis to nitrocellulose or nylon membranes for hybridization studies and other types of analyses has proven to be extremely useful. Such transfers of DNA to membranes are called Southern blots after E. M. Southern, who developed the technique.

Analysis of RNAs by Northern Blot Hybridizations

As DNA molecules can be transferred from agarose gels to nitrocellulose or nylon membranes for hybridization studies, RNA molecules can also be separated by agarose gel electrophoresis in a similar way and transferred for analyses analyzed. Such RNA transfers are used routinely in molecular genetics laboratories and are called *northern blots* as it is the mirror image of the Southern blotting technique.

The northern blot procedure is essentially identical to that used for Southern blot transfers however, degradation by RNA molecules are more sensitive to degradation by RNases. Thus, gloves must be worn at all times to prevent contamination of solutions with RNases on one's fingers. In addition, all glassware must be baked overnight to inactivate RNases, and all chemical reagents must be baked or be highly purified to ensure that they are RNase-free. RNase molecules are extremely stable enzymes; they are not inactivated by boiling or other treatments that would totally destroy most enzymes. Most RNA molecules contain considerable secondary structure. Therefore, they must be kept denatured during electrophoresis to separate them on the basis of size. Denaturation is accomplished by adding formaldehyde or some other chemical denaturant to the loading buffer and the electrophoresis running buffer. After transfer to an appropriate membrane, the RNA blot is hybridized to either RNA or DNA probes just as with Southern blots. Now an alternative to nitrocellulose, DBM or diazobenzyloxymethyl paper is used. It may be reused several times for different probes, without appreciable loss of nucleic acid signal.

Northern blot hybridizations are extremely useful in studies of gene expression. They can be used to determine whether a particular gene is transcribed in all tissues of an organism or only certain tissues. They can also be used to study the temporal pattern of expression of individual genes during growth and development. Major disadvantage with northern blot hybridizations is that it only measures the accumulation of RNA transcripts. but explains nothing about why the observed accumulation occurs.

Analysis of Proteins by Western Blot Techniques

Polyacrylamide gel electrophoresis has been used as an important tool for the separation and characterization of proteins.

Since many functional proteins are composed of two or more subunits, individual polypeptides are separated by electrophoresis in the presence of the detergent sodium dodecyl sulfate (SDS), which denatures the proteins (Fig. 3.10).

After electrophoresis, the proteins are commonly detected by treating with Coomassie blue or silver stain. However, the separated polypeptides in the gel can also be transferred to a nitrocellulose membrane, and individual proteins can be detected by using specific antibodies. This transfer of proteins from acrylamide gels to nitrocellulose membranes is called western blotting and is performed by using an electric current-so-called *electroblotting*. After transfer, a specific protein of interest is identified by placing the membrane with the immobilized proteins in a solution containing an antibody to the protein. Nonbound antibodies are then washed off the membrane, and the presence of the initial ("primary") antibody is detected by placing the membrane in a solution containing a "secondary" antibody. This "secondary" antibody reacts with immunoglobulins (the group of protein containing all antibodies) in general. The secondary antibody is conjugated to either a radioactive isotope (permitting autoradiography) or an enzyme that produces a visible product when the proper substrate is added. The western blot procedure is a very powerful tool for identifying and characterizing specific gene-products.

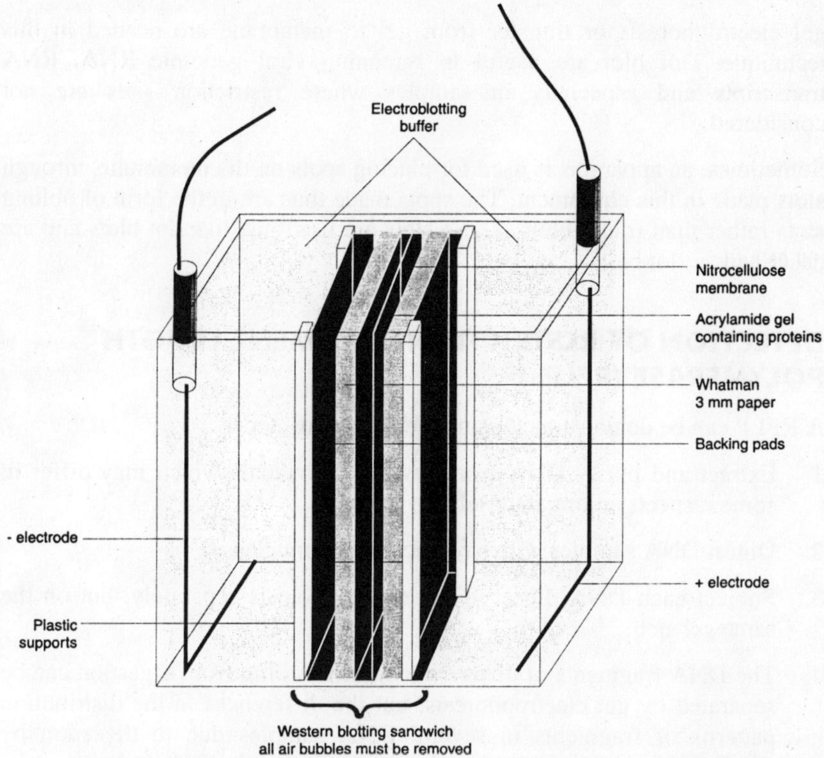

Fig. 3.10: Diagram of a typical Western blotting or electroblotting apparatus. An electric current is used to transfer the proteins from the gel to the nitrocellulose membrane placed next to it in the blotting sandwich. All other components of the sandwich function to provide gentle but firm support; tight contact between the gel and the membrane is essential for good transfer. All bubbles must be carefully removed from the sandwich to assure uniform flow of current and complete transfer of all proteins.

Dot blots and slot blots

Another variety of Southern or Northern blots is dot blots or slot blots. In dot blots, cloned or purified extract of DNA is tested by spotting them adjacent to each other on a nitrocellulose membrane. DNA blots thus produced are immobilized and denatured so that they are bound on the membrane as single stranded DNA blots. The membrane is then hybridized with radioactively labelled probe (DNA or RNA). The dot representing sequence related to probe will light up by autoradiography. The intensity of the dot will indicate the relative concentration of a sequence related to the probe in the DNA sample used for a particular dot. No steps involving digestion with enzyme, gel electrophoresis or transfer from gel to membrane are needed in this technique. Dot blot are useful in screening viral genomic RNA, RNA transcripts and especially in samples where restriction sites are not considered.

Sometimes, an apparatus is used for placing spots on the membrane, through slots made in this equipment. The spots made thus are in the form of oblong slots rather than round blots. These slots are used just like dot blots and are described as slot blots.

DETECTION OF RESTRICTION FRAGMENT LENGTH POLYMERASE (RFLPs)

A RFLP can be demonstrated using the following steps:

1. Extract and purify DNA from several individuals which may differ in some respects among themselves.
2. Digest DNA samples with a restriction endonuclease.
3. Subject each DNA digest to gel electrophoresis separately, but on the same gel slab.
4. The DNA fragments of different lengths resulting from digestion can be separated by gel electrophoresis, but the differences in the distribution patterns of fragments in several DNA samples due to digestion by several endonucleases can not be detected directly. This is because the number of fragments is large and the range in size is rather continuous, such that they form a continuous smear on the gel. Electrophoresis is, therefore, followed by the following steps.
5. Fragments are transferred from gel to nitrocellulose filters using the technique of **Southern blotting**.
6. The filters with small fragments, fully or partially homologous to the probe, can be detected by autoradiography after hybridization.

With a sufficient number of digests, it is possible to deduce the origin of each band on the gel, that is, which enzyme or combinations of enzymes have generated it. This yields a restriction enzyme cleavage map or restriction map. Obtaining such a map is very often essential first step in analyzing a piece of DNA of unknown structure. Another method is by labelling with probe followed by autoradiography. Since DNA clones having repetitive DNA may hybridize with many fragments, they will give a smear on autoradiography. Therefore, unique DNA sequences are generally used as probes for detecting RFLPs. The DNA fragments hybridizing with the probe, can be visualized by autoradiography, and will be called RFLPs at the phenotypic level. RFLPs can be studied in a set of related species, using a random or a specific DNA probe. The similarities and differences can be used to infer phylogenetic relationships. This has been done in a number of cases, both in plants and in animals. RFLPs has also been used to prepare chromosome maps in humans mice, fruitfly, maize, tomato, lettuce and rice.

AMPLIFIED RESTRICTION FRAGMENT LENGTH POLYMERASE (AFLP)

AFLP is a DNA fingerprinting technique that detects genomic restriction fragments and resembles in that respect the RFLP technique, with the major difference that PCR amplification instead of Southern hybridization is used for detection of restriction fragments.

AFLP technique was originally used to fingerprint genomic DNA sample (Zabeau and Vor, 1995). AFLP is based on selective amplification of a subset of restricted DNA fragments using the PCR technique. DNA is digested with a combination of two restriction endonucleases; double standard restriction-site-specific adopters are then ligated to the ends of digested DNA to generate templet for PCR amplification. The sequence of adapters in combination with restriction site sequence is used to design primers. To the primers are added selective nucleotides at the 3' end PCR leads to the amplification of only a selective set of DNA fragments where the nucleotides flanking the restriction sites match the selective nucleotide of the PCR primers. The primers can be radioactively labeled or one of the dNTPs during PCR is radio-labelled to facilitate aurodiography of fragments after electrophoresis on denaturing polyacrylamide gel. Each experiment, on an average yield about 50-100 bands in the size range of 50 bp to 350 bp,. The number of bands depends on the enzyme/primer combination.

DNA SEQUENCING

Restriction enzyme mapping and Southern blotting help to establish the gross organization of a piece of DNA, and allow the approximate position of sequences of interest to be determined. The next step in analysis is normally to determine the nucleotide sequence.

Methods for analyzing DNA sequence

Our present ability to sequence essentially any DNA molecule is the result of four major developments.

1. The most important breakthrough was the discovery of restriction enzymes and their use in preparing homogenous samples of specific segments of chromosomes.

2. A second major advance was the improvement of gel electrophoresis procedures to the point where DNA fragments that different in length by a single nucleotide could be resolved.
3. The development of gene cloning techniques greatly facilitated the preparation of large quantities of a particular gene or DNA sequence of interest for use in sequencing studies.
4. Finally, two different approaches to the rapid sequencing of DNA molecules (e.i. by chemical degradation and by chain termination) were developed in the period from 1974 to 1977.

Sequencing by chemical degradation

Determination of DNA sequence depends upon the resolution obtained from polycylamide gels, which are run in the presence of denaturant e.g. urea. If two single-stranded DNA molecules, identical in sequence over most of their length but differing in just one base on one of the two molecules, are subjected to electrophoresis they will be separated. In the chemical degradation procedure a DNA restriction fragment is isotopically labelled with P at either its 5' or its 3' termini and each of the two strands is separated and isolated.

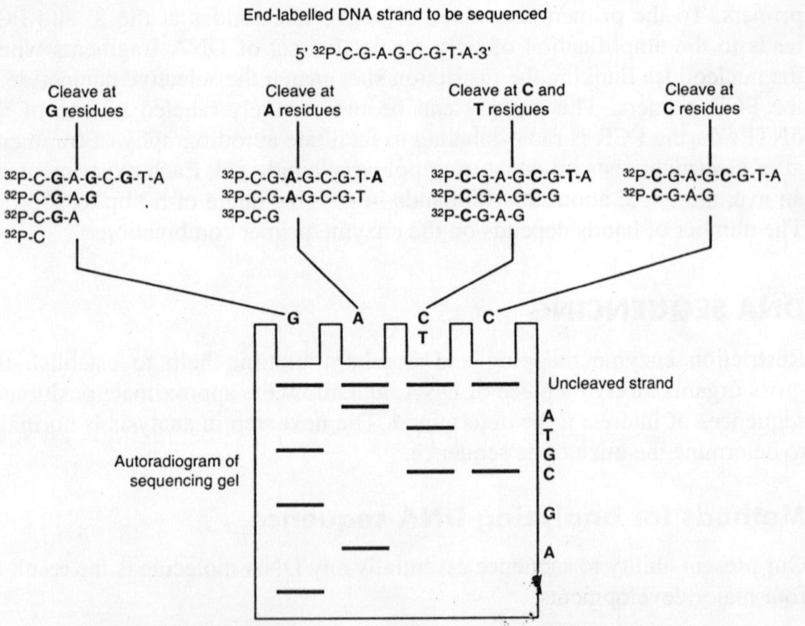

Fig. 3.11: DNA sequen0ce analysis by the chemical degradation (Maxam and Gilbert) technique.

The DNA is then partially modified with chemical reagents specific for the different bases, and cleaved at the modified nucleotides. This generates a set of molecules differing in length but with the same isotopically labelled terminus. These fragments are subjected to denaturing gel electrophoresis in parallel slots of high resolution gel, and the sequence of the DNA can be deduced from the resulting autoradiogram (Fig. 3.11).

Sequencing by chain termination

In practice the chemical degradation procedure has been replaced by chain termination method. In this procedure DNA is synthesized rather than degraded, but the principle is similar to the chemical method. DNA polymerases function by adding a primer, a short synthetic oligonucleotide, to pre-existent DNA strand, to provide a fixed 5' terminus for the growing chain.

The reaction is performed in the presence of chain terminators. These nucleotide derivatives act as blocking group at their 3' hydroxyl terminal and once these are incorporated into the growing chain, they prevent further addition of nucleotides. For example, if dideoxy GTP (Fig. 3.12), is mixed with unmodified GTP and the other three nucleoside triphosphates, terminates at a subset of G residues in the newly synthesized strand (Fig. 3.12). Four parallel reactions are run, with dideoxy radiolabelled nucleoside triphosphates of all four kinds are included. The nucleotide sequence of the template strand can be deduced by comparison of the pattern of autoradiographic bands in the four lanes of the gel (Fig. 3.12, 3.13).

A suitable template for single-strand DNA sequence analysis can be created by inserting the double-stranded template DNA into the genome of a bacterial virus, which naturally packages only one of the two strands into viral particles. Another method is denaturing of double-stranded DNA template by treatment with alkali. An excess of primer is added before re-annealing occurs, so that primer will compete with the complementary strand for annealing to the template. This is a preferable method over insertion of DNA into viral vector.

DNA sequence analysis using chain terminators is best performed using a single-stranded template. This allows the primer to be free accessible to the template.

The major disadvantage of double-stranded DNA sequence analysis is the presence of competing complementary strand. This leads to a higher level of sequencing artefacts. Like presence of more than one band. In such situation it may be impossible to decide which is the correct residue at that position. Artefacts of this kind are not restricted to double-strand DNA sequence analysis; some level of artefacts is a universal problem in sequence analysis.

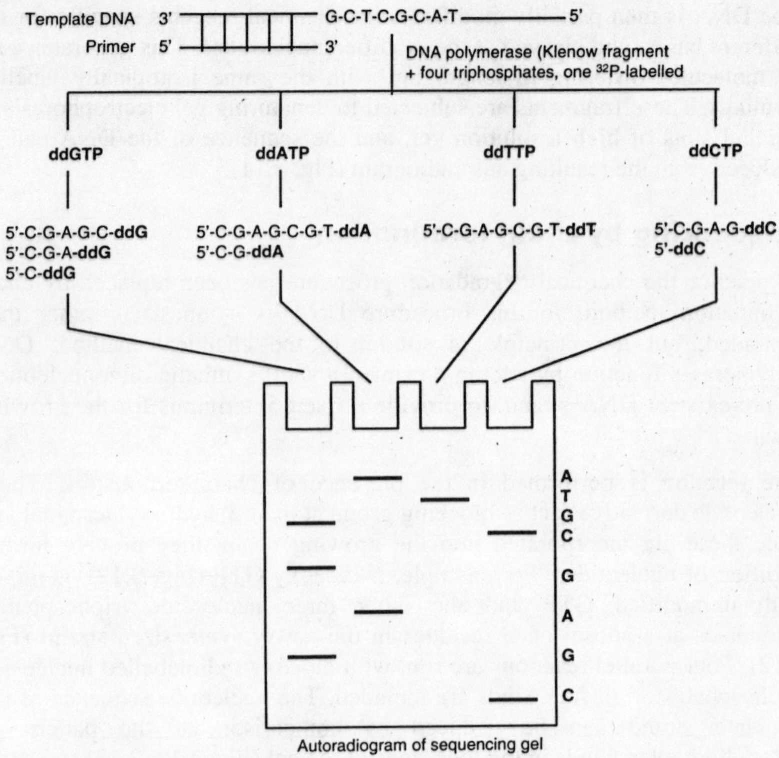

Fig. 3.12: DNA sequence analysis by the chain termination (Sanger) method.

The DNA sample then re-sequenced using another primer, or by using the other strand (i.e. the strand which is complementary to that used in the original analysis) as template.

Even when a single-stranded DNA, molecule with more than 500 nucleotides, clearly it is not possible on a polyacrylamide gel to resolve when the molecules are larger than about the nucleotides. This sets the upper limit to the length of sequence to be determined.

One way to extend from a known into an unknown region of sequence is to synthesize an oligonucleotide primer which is complementary to a tract near the end of the region whose sequence has already been established. In this way, it is possible to 'walk' along a sequence, at each step synthesizing a new primer which is used to extend the region of known sequence (Figure 3.14). Alternatively, it is possible to delete sequences enzymatically from the template at a position downstream of the annealing site of the primer, to generate a deletion series. The nucleotide sequence of a set of such molecules can be used to deduce the sequence of the intact template.

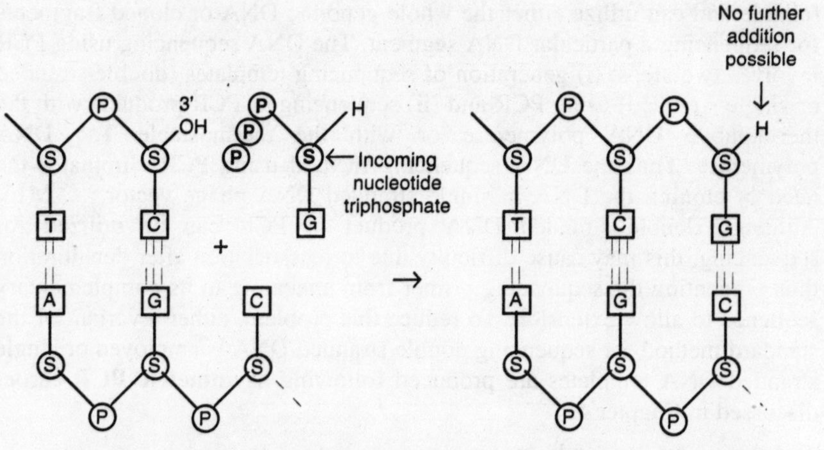

Fig. 3.13: The chemical basis of sequence analysis using chain terminators. This is a representation of a reaction in which a dideoxy G residue incorporated into the growing chain blocks further incorporation.

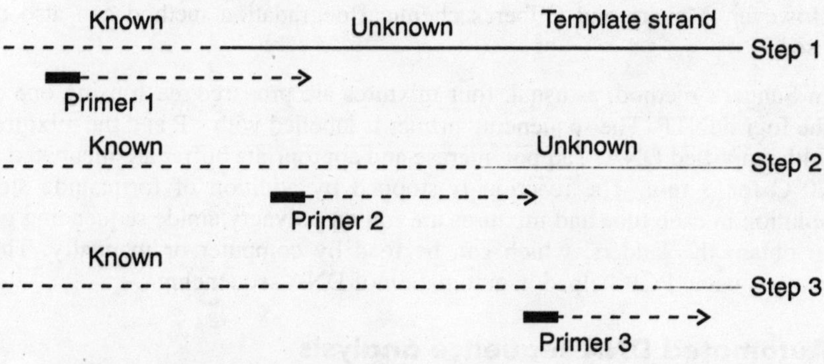

Fig. 3.14: DNA sequence analysis by *primer walking.*

By determining the sequence of both strands of a DNA molecule, and confirming that the two deduced sequences are perfectly complementary, it is possible to detect and eliminate artefacts generated in the reactions or during electrophoresis. When this check is performed, DNA sequence can be established with a very high degree of accuracy.

Direct DNA sequencing using PCR (also called ligation mediated PCR=LMPCR)

Polymerase chain reaction (PCR) has also been used for sequencing the amplified DNA product. This method of DNA sequencing is faster and more reliable and can utilize either the whole genomic DNA or cloned fragments for sequencing a particular DNA segment. The DNA sequencing using PCR involves two steps: (i) generation of sequencing templates (double stranded or single stranded) using PCR and (ii) sequencing of PCR products with the thermolabile DNA polymerase or with the thermostable Taq DNA polymerase. Thus the DNA sequencing method using PCR eliminates the need of cloning the DNA in single stranded DNA phage vector i.e. M13. Although double stranded DNA product of PCR can be utilized for sequencing, this may cause difficulty due to reassociation after denaturation thus preventing the sequencing primer from annealing to its complementary sequence to allow extension. To reduce this problem, either a variant of the standard method for sequencing double stranded DNA is employed or single stranded DNA templates are produced following **asymmetric PCR** earlier discussed in Chapter 4.

A number of thermolabile DNA polymerases have been used for sequencing of *in vitro* amplified DNA. Alternatively, **Taq polymerase** (thermostable enzyme) used for PCR, can also be used for sequencing reaction. In either case, Sanger's synthetic method involving incorporation of dideoxynucleotide (ddNTP) for chain termination, is used for sequencing. However, Maxam and Gilbert's chemical degradation method can also be used.

In Sanger's method, as usual, four mixtures are prepared, each using one of the four ddNTP. The sequencing primer is labelled with ^{32}P and the mixtures with amplified DNA, Taq polymerase and appropriate buffer are incubated at 70°C for 5 min. The reaction is stopped by addition of formamide stop solution in each tube and mixtures are run on polyacrylamide sequencing gel to obtain the ladders, which can be read by computer or manually. This method using PCR helped in automation of DNA sequencing.

Automated DNA sequence analysis

Using the chain termination method, DNA sequencing can be mechanized to make the process rapid. In a conventional sequence reaction described above four separate single-strand extension reactions are performed, with each containing a different chain-terminating dideoxynucleoside triphosphate. The radiolabelled reaction products are separated on four separate tracks of a polyacrylamide gel. In machine-based sequencing a single reaction is run, with each of the four dideoxynucleoside triphosphates containing a different fluorescent label.

When excited by an argon laser, the fluorescent labels emit light at a different wavelength. The same chain-termination reactions are carried out; however, it is no longer necessary to run the products in separate lanes of a gel. The reaction products are mixed and run in a single lane in a gel and as the different length fragments pass the laser each fluorescent label gives a unique signal which is stored in the computer.

Machine-based DNA sequencing is beginning to change the speed and scope of sequencing projects dramatically. The first complete genomic sequences were those of simple viruses such as the SV40 tumor virus, containing around 500 based of DNA. Recently, the DNA sequence of over 106 000 bases of DNA from human chromosome 19 was determined in a few months using DNA sequencing machines. The sequence of a significant fraction of the *E.coli* genome has been established, the entire sequence of yeast chromosome 3 is known, and there are plans eventually to determine the sequence of the entire human genome.

MECHANICAL SHEARING OF DNA

In addition to digestion of DNA with restriction endonucleases to produce discrete fragments, there are a variety of treatments which result in non specific breakage. Non specific endonucleases and chemical degradation can be used but the only method that has been much applied to gene manipulation involves mechanical shearing.

The long, thin threads which constitute duplex DNA can not be easily broken by shear forces in solution. Intense sonication with ultrasound can reduce the length to about 300 nucleotide pairs. More controlled shearing can be achieved by high-speed stirring in a blender. Breakage occurs essentially at random with respect to DNA sequence. The termini consist of short, single-stranded regions which may have to be taken into account in subsequent joining procedures.

TRANSCRIPT MAPPING TECHNIQUES

Analyzing the nucleotide sequence of a piece of DNA is a necessary preliminary to understand its precise organization and function, but is not normally sufficient for these purposes. Because eukaryotic gene transcripts are subjected to a complex series of post-transcriptional processing events, the structure of a gene can be understood only by correlating the organization of the genomic DNA with the organization of its mRNA transcript. This has required the establishment of sophisticated cloning and mapping techniques which can, when used in conjunction, give a complete description of the structure of a gene.

It is possible to isolate, by gene cloning, the double-stranded DNA copies of individual mRNA sequences. These cDNA clones will normally contain a DNA copy of the poly(A) sequence at the 3' terminus. Hence, by determining which strand of the cDNA clone contains a poly(A) tract at one end, it is possible to identify the coding strand of the cDNA clone. Provided a sufficient length of sequence has been established to make their presence statistically likely (in practice 300-500 nucleotides), termination codons will occur in two of the three potential coding frames, ruling them out as translational open reading frames (ORFs).

Sometimes this level of analysis, that is, identification of the reading frame and deduction of the encoded protein sequence, will be all that is required. However, it is often desirable to determine the structure of the gene itself. It will have been isolated as part of a genomic clone. This will contain the gene and a variable amount of flanking DNA depending upon the method of cloning used.

By correlating the sequence of the gene with the sequence of the mRNA deduced from the cDNA clone, it is possible to identify the site of poly(A) addition and the positions of introns. However, technical limitations in the cDNA cloning procedure ensure that the cDNA clone will not normally contain sequences derived from the extreme 5' terminus of the mRNA. There are two potential methods to determine the position of the 5' end, primer extension and S1 mapping.

Primer extension

In primer extension, a short oligonucleotide complementary to a sequence tract near to the 5' end of the mRNA is annealed to the mRNA. This is then used to prime synthesis with the enzyme reverse transcriptase (Fig. 3.16). This enzyme produces a DNA copy of the mRNA (a cDNA) which extends to the 5' terminus of the mRNA. By correlating the length of the cDNA primer extension product with the sequence of the genomic DNA, it is possible to deduce the start site of transcription.

S1 mapping

The alternative technique is called S1 mapping (Fig. 3.17). An isotopically labelled probe is generated from the genomic clone. This probe is complementary to the mRNA over a region encompassing part or all of the first exon, and extending upstream of the region where the cap site is expected to be located. The cap site may have already been putatively localized using primer extension. The probe is annealed to the mRNA and digested with an enzyme which destroys single-stranded nucleic acids but leaves double-stranded molecules intact. In early experiments an enzyme called nuclease S1 was generally used, and the name S1 mapping has been retained, even though other nucleases are now sometimes used. During digestion, the 3' proximal region of the probe, which has no complement in the mRNA molecule and which therefore remains single-stranded, is degraded. A molecule is generated which is the complement of the region between the cap site and the position at which the 5' terminus of the probe annealed to the mRNA. Again, by determining the length of this molecule it is possible to infer the position of the cap site.

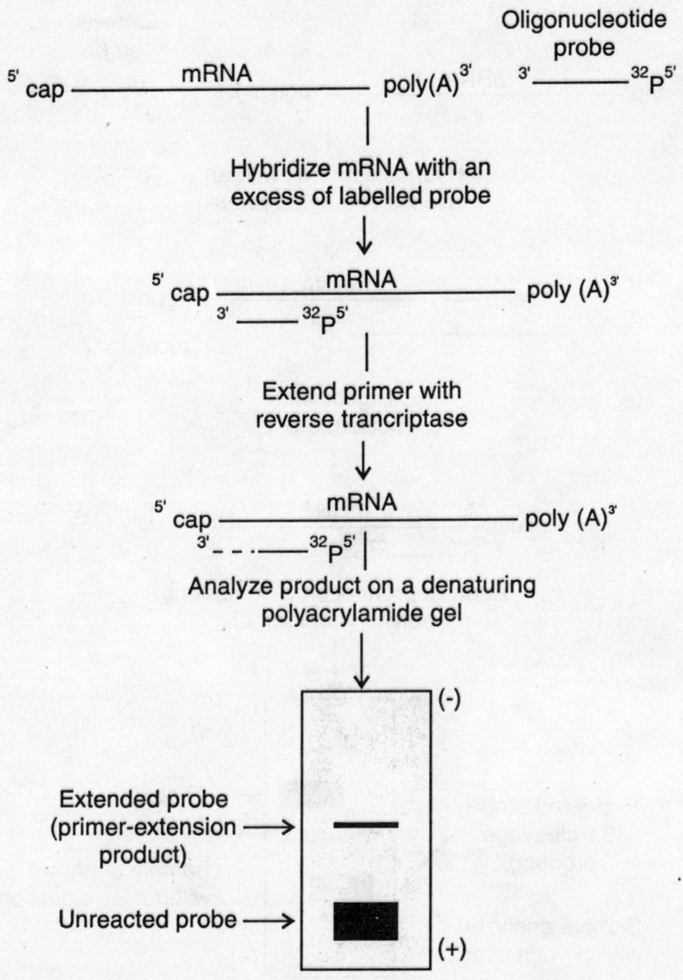

Fig. 3.15: Identification of the 5' end of an mRNA by primer extension.

These two techniques are often used in parallel to determine the cap site. They are both subject to artefacts, but of different kinds. Therefore, if both methods agree, it greatly increases confidence in the assignment of the cap site. If the hybridization is performed with an excess of probe, the amount of primer extension or nuclease-resistant product is proportional to the amount of primer extension or nuclease-resistant product is proportional to the amount of complementary mRNA. Both techniques can therefore be used an alternative to Northern transfer to quantitate mRNA levels.

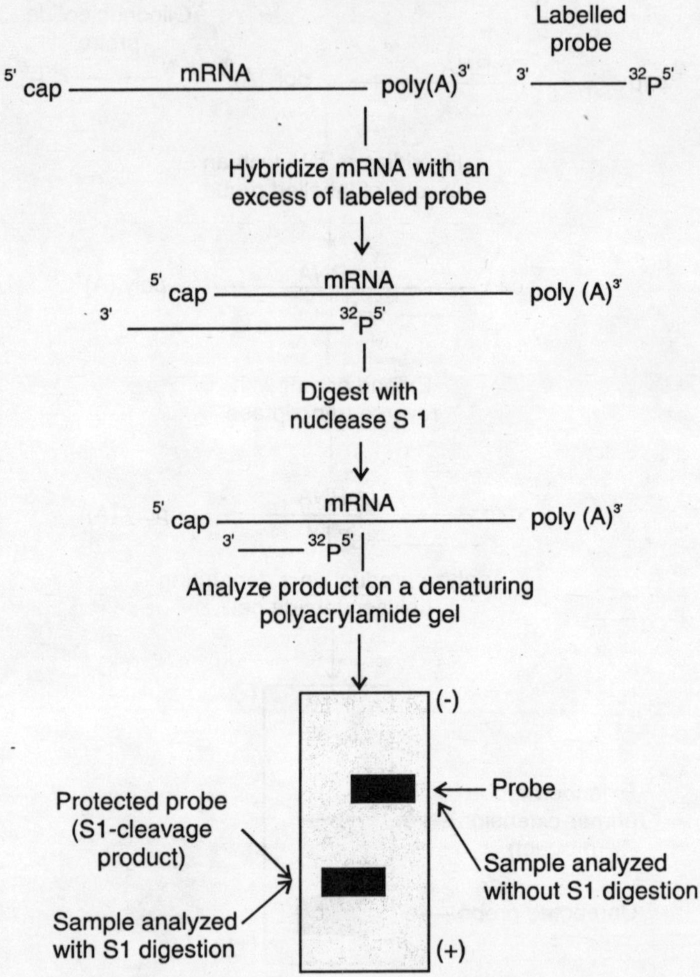

Fig. 3.16: Identification of the 5' end an mRNA by S1 mapping.

SYNTHESIS OF cDNA FROM mRNA

Complementary DNA can be prepared by isolating mRNAs from tissues which are actively synthesizing proteins, like roots and leaves in plants, ovaries or recticulocytes in mammals etc. Isolated and purified mRNA can then be directly used to synthesise complementary DNA strand using the enzyme reverse transcriptase. This results in a DNA/RNA hybrid. The RNA part of the hybrid is then removed with alkali. Thus a single-stranded DNA is generated which can act as the template for the synthesis of its complementary strand in the presence of DNA polymerase. The double stranded DNA molecule thus formed is called complementary DNA (cDNA), as it is synthesised complementary to the mRNA. In the prokaryotes the double stranded cDNA will be identical to the gene (DNA) from which the mRNA is formed. But, most of the eukaryotic genes contain intervening sequences (introns) which interrupt the coding sequences. Hence, the cDNA of the eukaryotic cells and their native DNA are not colinear.

As bacterial genes do not have the intron - exon organisation found in the eukaryotic genes they find the expression of eukaryotic genes in them (bacterial) difficult. Therefore one has to depend upon cDNA of eukaryotes, which does not have introns, for cloning.

Preparation of Double-Stranded cDNA

The mRNA that is specific for a particular kind of protein is isolated and purified. All naturally occurring mRNAs of eukaryotes have about 200 adenylate residues at their 3'-termini. This is called polyadenylate tail. The enzyme reverse transcriptase can mediate the synthesis of a new DNA strand only when it has a growing point in the form of a pre-existing primer, which is base paired with the mRNA template and has a free 3' OH group. Such a primer is provided by hybridizing a short complementary oligo (dT) molecule with the poly adenylate (poly-A) sequence.

The 3'OH end of the poly-T primer is made double standard by reverse transcriptase forming RNA-DNA hybrid. The RNA strand of this RNA - cDNA hybrid is digested by alkaline hydrolysis (DNA is resistant to alkali). At the 3'-OH end of the newly formed single stranded cDNA, a hairpin is formed which is stabilised by base - pairing. This hairpin which has a 3'-OH free end acts as a self-priming structure. Forming second strand of the cDNA in the presence of the fragment of the *E. coli* DNA polymerase-1 (Kornberg enzyme) and the four deoxynucleotide triphosphates.

The enzyme reverse transcriptase or T4DNA polymerase can also be used for the synthesis of the second strand of cDNA. The synthesised double strandede DNA is treated with the single-strand specific nuclease S1. This enzyme digests the curved region of the hair pin and any polynucleotide tails. Thus, finally a double stranded cDNA molecule is formed. This DNA can now be inserted into a suitable vector and can be further cloned in the desired host.

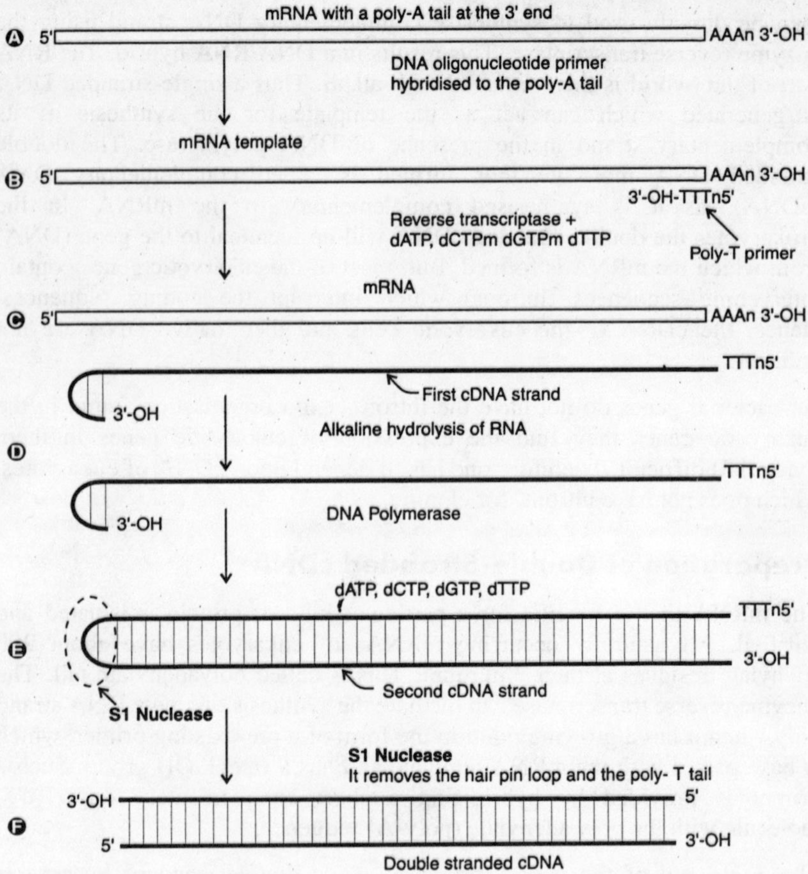

Fig. 3.17: Synthesis of double stranded cDNA from an mRNA.

Complementary DNA of a given mRNA is inserted into a suitable vector can then be cloned in a host. The cDNA is blunt ended. It has no cohesive ends that are complementary to the cohesive ends of any of the vectors. This is overcome by homopolymer tailing. Once inside the host cell, the cDNA in the hybrid will express the gene(s) it carries.

cDNA Library

Complementary DNAs (cDNAs) of the genome of an organism can be synthesised and cloned to generate a cDNA library. Thus, cDNA library is a collection of cDNA clones of an organism. The starting material for the synthesis of cDNA is the mRNA. Hence the resulting cDNAs will represent the mRNAs present in the cells from where these mRNAs were isolated. In the case of multicellular organisms there is specialisation of individual cells. For example, a mammal is made up of a large number of cell types: brain cells, mammary cells, blood cells, liver cells, cells of the pancreas, nerve cells etc. Each type of cell will contain the same complement of genes. However, only a particular set of genes will be expressed in a given type of cell. Other genes will remain silent. Only those genes that are expressed are transcribed into mRNA. Hence the cDNA clones of this type of cell will be only a selection of the total number of genes in a cell. Hence, for such organisms, cDNA clones of different cells are prepared.

SEARCHING FOR GENE USING COMPUTERS

Although the series of analyses described above are necessary ultimately to establish the organization and mode of expression of a particular gene, genes can be sought by less laborious means using computer programs. These search for regions of a DNA sequence which look like coding regions flanked by the signals that direct gene transcription and processing of their RNA products. This procedure is of greatest value when large regions of genomic DNA for which cDNA clones are not available are being analyzed. It is thus an essential part of the Human Genome Project.

One of the principles, that of searching for long ORFs (= open reading frames), has already been described. Although the sequence (the Kozak sequence) that precedes the ATG initiation codon is only weakly conserved between different genes. It can provide a helpful guide as to the start site of translation. The start site of transcription often lies downstream of a TATA box, and it is also possible to search for the consensus binding sites of other transcription factors.

RNA processing signals can also be used to establish the likely gene organization. The sequence that signals polyadenylation is well conserved, and splice junctions have a consensus sequence comprising an almost perfectly conserved core (5' GU----AG 3'), with moderately well conserved flanking sequences.

Identification genes by analysis of DNA sequences is an evolving science. The structures of new genes are constantly adding and the sophistication of the search programs is increasing. The databanks exists which contain many thousands of gene sequences e.g. Gene bank and the EMBL. Although some scientists initially submit their gene sequence data to one rather than other, the two organizations that compile the databases then exchange data.

4

Techniques of Producing Transgenic Plants

Three gene-transfer techniques have been developed to such a stage of reproducibility and practical applicability that they can be recommended for future practical work. These are *Agrobacterium-mediated* gene transfer, *ballistic* gene transfer, and *direct* gene transfer to *protoplasts.* These techniques are described in some detail in the following chapter, the Table 4.1 summarizes which crop plants have been transformed with which of these techniques. These are additional techniques that have yielded confirmed transgenic plants and that show some promise for future development, such as *microtargeting, microinjection, electroporation* into tissues, and *whiskers.* These techniques are described to a smaller extent and the literature citations will support further reading. There is a further group of techniques for which unequivocal evidence for the production of transgenic plants is still missing such as the pollen tube-pathway, and for which the reader is referred to a more detailed assessment (Potrykus, I.).

Table 4.1: Transgenic Crop Plants Obtained by *Agrobacterium*-Mediated Transformation.

Latin name	English name	Method	Ref.
Medicago sativa	Alfalfa	A	(Schroeder et al., 1998), (D'Halluin et al., 1990)
Malus pumila	Apple	A	(James et al., 1989), (Lambert and Tepfer 1991)
Prunus armeniaca	Apricot	A	(Laimer et al., 1992)
Asparagus officinalis	Asparagus	A	(Delbreil et al., 1993)
Populus tremula	Aspen	A	(Block 1990), (Tsai et al., 1994)
Hordeum vulgare	Barley	B	(Wan and Lemaux 1994)-(Jähne et al., 1994)
Phaseolus vulgaris	Bean	B	(Russell et al., 1993)
Lotus corniculatus	Birdsfot trefoil	A	(Webb et al., 1990)

Latin name	English name	Method	Ref.
Ribes nigrum	Black currant	A	(Graham and McNicol 1991)
Dianthus sp. caryophyllus	Carnation	A	(Nugent et al., 1991)
Citrus	Carrizo Citrange	A	(Moore et al., 1992)
Daucus carota	Carrot	A	(Scott and Draper 1987)
Brassica oleracea	Cualiflower	A	(Block et al., 1989)
Apium graveoleus	Celery	A	(Catlin et al., 1988)
Cicer arietinum	Chickpea	A	(Fontana et al., 1993)
Cichorium endivia	Chicory	A	(Vermeulen et al., 1992)
Chrysanthemum sp.	Chrysanthemum	A	(Ledger et al., 1991)
Gossypium hisutum	Cotton		(Firoozabady et al., 1987)
		A	(Finer and McMullen 1990),
		B	(McCade and Martinell 1993)
Zea mays	Corn	P	(Omirulleh et al., 1993)
		B	(Gordan-Kamm et al., 1990), (Vain et al., 1993), (Koziel et al., 1993)
Vaccinium macrocarpon	Cranberry	B	(Serres et al., 1992)
Cucumis sativus	Cucumber	A	(Chee 1990)
Festuca rubra	Fescue, red	P	(Spangenberg et al., 1994)
		B	(Spangenberg et al., 1995)
F. arundinacea	Fescue, tall	P B	(Wang et al., 1992) (Spangenberg et al., 1995)
Linum usitatissimum	Flax	A	(Jordan and McHughen 1987), (McHughen 1989)
Gerbera hybrida	Gerbera	A	(Elomaa et al., 1993)
Vitis spp.	Grapevine	A	(Baribault et al., 1990), (Torregrosa et al., 1994)
Actinidia chinensis	Kiwi fruit	A	(Uematsu et al., 1991), (Rugini et al., 1991)
Lactuca sativa	Lettuce	A	(Michelmore et al., 1987)
Cucumis melo	Muskmelon	A	(Gang and Grumet 1990), Dong et al., 1991)
Brassica juncea	Mustard	A	(Barfield and Pua 1991)
Avena sativa	Oat	B	(Somers et al., 1992)
Brassica napus	Oilseed rape (canola)	B	(Chen and Beversdorf 1994)
		A	(Block et al., 1989), (Radke et al., 1998)
Dactylis glomerata	Orchardgrass	P	(Horn et al., 1998)
Dendrobium sp.	Orchid	B	(Chia et al., 1994)
Carica papaya	Papaya	A B	(Fitch et al., 1993) (Manshardt et al., 1992)
Passiflora edulis	Passionfruit	A	(Manders et al., 1994)
Pisum sativum	Pea	A	(Puonti-Kaerlas et al., 1990)-((Schroeder et al., 1993)
Arachis Hypogaea	Peanut	A	(Eapen and George 1994)
		B	(Brar et al., 1994)
Carya ovata	Pecan	A	(McGranahan et al., 1993)
Lolium perenne	Penennial ryegrass		(Spangenberg et al., 1995)
Solanum muricatum	Pepino	A	(Atkinson and Gardner 1991)
Petunia hybrida	Petunia	P	(Krüger-Lebus and Potrykus 1987)

Latin name	English name	Method	Ref.
Prunus domestica	Plum	A	(Mante et al.,1991), (Scorza et al., 1994)
Populus sp.	Poplar	A	(Filatti et al., 1987)
		B	(McCrown et al., 1991), (Wilde et al., 1992)
Solanum tuberosum	Potato	A	(Sheerman and Bevan 1998), (Stiekema et al., 1998)
		P	(Masson et al., 1989)
Boehmeria nivea	Ramie	A	(Dshi et al., 1993)
Oryza sativa	Rice	A	(Hiei et al., 1994)
		P	(Shimamoto et al., 1989), (Datta et al., 1990)
		B	(Christou et al., 1991), (Li et al., 1993)
Secale cereale	Rye	B	(Castillo et al., 1994)
Sorghum bicolor	Sorghum	B	(Casas et al., 1993)
Glycine max	Soybean	A	(Hinchee et al., 1998), (Chee et al., 1989)
		B	(McCabe et al., 1988)
Fragaria x Anannassa	Strawberry	A	(James et al., 1990), (Nehra et al., 1990)
Stylosanthes humilis	Stylo	A	(Mammers 1988)
Trifolium subterraneum	Subterranean clover	A	(Khan et al., 1994)
Beta vulgaris	Sugar beet	A	(Lindsey and Gallois 1990)
Saccharum officinarum	Sugarcane	B	(Bower and Birch 1992)
Helianthus annuus	Sunflower	A B	(Everett et al., 1987)-(Knittel et al., 1994) (Bidney et al., 1992)
Liquidambar styraciflua	Sweetgum	A	(Sullivan and Lagrimini 1993)
Cyphomandra betacea	Tamarillo	A	(Atkinson and Gardner 1993)
Physalis ixocarpa	Tomatillo	A	(Assad-Gracia et al., 1992)
Lycopersicon esculentum	Tomato	A	(Fillatti et al., 1987), (Shahin et al., 1986)
Poncirus trifoliata	Trifoliate orange	A	(Kaneyoshi et al., 1994)
Hordeum x Triticum	Tritordeum	B	(Barcelo et al., 1994)
Nicotiana tobacum	Tobacco	A	(Horsch et al., 1984)
		P	(Paszkowski et al., 1984)
		B	(Tomes et al., 1990)
Agrostis alba, Agrostis palustris	Turf grasses: creeping bentgrass; Redtop	P B	(Asano and Ugaki 1994) (Hartman et al., 1994)
Juglans regia	Walnut	A	(McGranahan et al., 1988), (Leslie et al., 1990)
Citrullus vulgaris	Watermelon	A	(Choi et al., 1994)
Triticum aestivum	Wheat	P	(He et al., 1994)
		B	(Becker et al., 1994), (Vasil et al., 1992)-(Nehra et al., 1994)
Trifolium repens	White clover	A	(White and Greenwood 1987), (Voisey et al., 1994)

Latin name	English name	Method	Ref.
Sinapis alba	White mustard	A	(Hadfi and Batschauer 1994)
Picea glauca	White spruce	B	(Ellis et al., 1993)

(A) = *Agrobacterium* mediated transformation; (B) = Ballistic Methods; (P) = or Direct Gene Transfer to Protoplasts.

Table 4.2: References on Selectable Marker Genes Used in Transformation of Crop Plants

Species	*npt2*	*hpt*	*Bar*
Wheat	(Nehra et al., 1994)	-	(Vasil et al., 1992).
Barley	(Ritals et al., 1994)	-	(Wan and Lemaux 1994)
Maize	(D'Halluin et al., 1992)	(Walters et al., 1992)	(Gordon-Komm et al., 1990)
Rice (japonica)	(Battraw and Hall 1992)	(Shimamoto et al., 1989)	(Christou et al., 1991)
Rice (indica)	(Peng et al., 1992)	(Datta et al., 1990)	(Datta et al., 1992)
Oat	-	-	(Somers et al., 1992)
Sorghum	-	-	(Casas et al., 1993)
Sugarcane	(Bower and Birch 1992)	-	
Soybean	(McCabe et al., 1988) (Hinchee et al., 1988)	(Finer and McMullen 1991)	
Pea	(Davies et al., 1993)	(Puonti-Kaerlas et al., 1990)	(Schroeder et al., 1993)
Peanut	(Mansur et al., 1993)	(Ozias-Akins et al., 1993)	(Brar et al., 1993)
Alfalfa	(Shahin et al., 1986)	(D'Halluin et al., 1990)	(D'Halluin et al., 1990)
Rapessed	(Fry et al., 1987)	-	(Block et al., 1987)
Arabidopsis	(Valvekens et al., 1988)	(Llyod et al., 1986)	(Sawasaki et al., 1994)
Cotton	(Firoozabady et al., 1987)	(Finer and McMullen 1990)	
Sugar beet	(Lindsey and Gallois 1990)	-	(Denecke et al., 1992)
Sunflower	(Schrammeijer et al., 1990)	-	
Potato	(Shahin and Simpson 1986)	-	(De Block et al., 1987)
Tobacco	(Horsch et al., 1984)	(Waldron et al., 1985)	(De Block et al., 1987)
Tomato	(Shahin et al., 1986)	-	(De Block et al., 1987)
Lettuce	(Michelmore et al., 1987)	-	
Carrot	(Pawlicki et al., 1992)	-	

Species	npt2	hpt	Bar
Cucumber	(Trulson et al., 1986)	-	
Melon	(Fang and Grumet et al., 1990)	-	
Apple	(James et al., 1989)	-	
Citrus	(Moore et al., 1992)	-	
Poplar	(Klopfenstein et al., 1991)	-	(De Block et al., 1990)

Several approaches have been used to produce transgenic plants. The areas follows:

AGROBACTERIUM MEDIATED GENE TRANSFER

For genetic transformation in plants *Agrobacterium tumefaciens* (a soil bacterium) is used. It transfers T-DNA into plant genome. Foreign genes are inserted in T-DNA through *Ti*-plasmid, co-transferred and integrated in the host genome. Only some sequences of T-DNA in *cis*-position are required for its transfer. By deleting T-DNA encoded genes and adding selectable markers a *disarmed plasmid vector* is prepared without disturbing endogenous balance. Disarmed strains with *Ti*-plasmid can be used to produce transgenic plants.

Plant-transformation Vectors

These vectors are designed on the basis of naturally occurring gene-transfer mechanisms. They are widely employed in plant molecular biology and in the genetic engineering of plants (Fraley et al., 1985, 1986; Hiel et al, 1997; Tyagi et al., 1999). Chilton (1993) described these vectors as *poor man's vector.* They are used for gene delivery to higher plants.

Advantages of Vector

1. They are convenient and efficient.
2. They allows the transfer of large segments of DNA with minimal rearrangements.
3. They are the integration of small numbers of copies of genes into plant chromosomes (Hamilton et al., 1996; Klee et al., 1987; Zambryski, 1992).
4. Foreign genes delivered by this method are usually transmitted to progeny plants in a Mendelian manners.
5. In the *Ti*-plasmid, the genes responsible for the transfer of T-DNA are clustered together in a region. The transfer process is active even when the virulence genes and the T-DNA are located on separate replicon in an *A. tumefaciens* cell.

6. In the widely used binary system, a *Ti*-plasmid services as helper; providing the virulence functions, and artificial T-DNA, which contains a selectable marker gene and other genes of interest, is placed on a second, small plasmid, which is often referred to as binary vector (An et al., 1998).
7. In dicotyledonous plants the systems of *Agrobacterium* mediated transformation have been well established for many. However, monocotyledonous plants (cereal/grasses) were originally outside, the host range of *A. tumefaciens.*

GENE CLONING

Gene cloning is a technique, where an isolated single copy of a gene is cloned to obtain its indefinite identical copies. In gene cloning, we introduce a vector containing the desired gene into the host cell. Cloning can be done in bacterial host (e.g. *E. coli*) or in eukaryotes (e.g. yeast). Basically, cloning involves four steps:

1. The vector DNA is cleaved with one or more restriction enzymes;
2. The DNA to be cloned, the target or insert is joined to the vector, generating a recombinant molecule;
3. The recombinant DNA molecules are introduced into the host bacterial cell. The host cell is said to be *transformed* by the introduction of such vector molecule;
4. Transformed colony is selected and amplified.

Another method of amplifying the desired gene is PCR, which offers this possibility in test tube. The two methods PCR and gene cloning are complementary techniques. Molecular cloning of a PCR product is now commonly used to isolate and manipulate genes. In this chapter, we will discuss gene cloning in *E. coli* and yeast, as well as PCR.

Creating Recombinant DNA

Recombinant DNA molecules are produced by inserting a foreign DNA segment into the DNA molecule of a vector. This involves joining of the broken ends of vector DNA with the two ends of sequence to be cloned. These recombinant DNA molecules (also called as rDNA molecule) are subsequently cloned. There are three methods of joining fragments *in vitro.* The first uses ability of DNA ligase to join the cohesive ends produced by certain restriction enzymes. The second depends on the ability of T4 ligase to form phosphodiester bonds between blunt-ended fragments. The third utilizes the enzyme terminal deoxynucleotidyl transferase to synthesize homopolymeric tails at the ends of fragments. The three enzymes and their mechanism of joining DNA are separately dealt below.

DNA Ligase

The sticky ends created by restriction enzyme has the nick after a few base pairs in opposite strands. DNA ligase repairs these nicks and forms intact duplex (also see chapter 3). The optimum temperature for ligation of nicked DNA is 37°C, but at this temperature the hydrogen-bonded joint between the sticky ends is unstable. Thus temperature is kept in range of 4 to 15°C.

This technique has the advantage of regenerating two restriction sites in recombinant DNA. Thus, the foreign DNA segment can be retrieved from cloned DNA copies by cleaving again with the same enzyme. The disadvantages within this technique include:

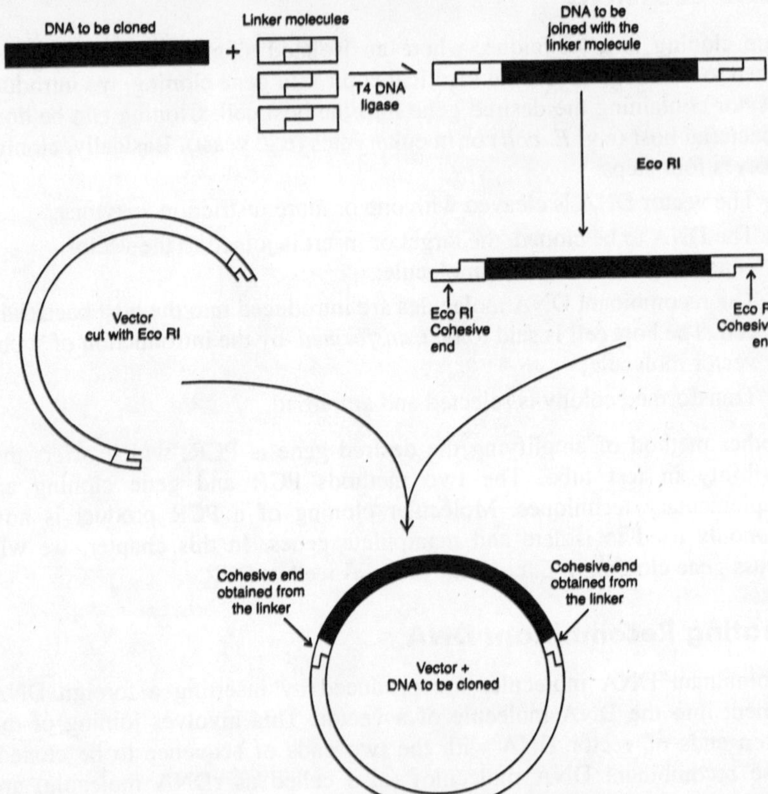

Fig. 4.1: Linker molecules containing Eco RI cleavage site are joined by T4 DNA ligase to both the ends of blunt-ended DNA-to-be-cloned. Cohesive ends are then generated by cutting with Eco RI. This DNA is then inserted into a vector molecule cut by the same restriction enzyme.

1. The cohesive ends of foreign DNA may join end to end and self-circularize before getting inserted in vector.
2. The recognition site, particularly in sequence to be cloned, may not lie at a convenient position so that sometimes only a part of gene is inserted.

T4 DNA ligase

The *E.coli* DNA ligase can not catalyse ligation of blunt ends except under special reaction conditions of macromolecular crowding. Such blunt end ligation can be done with T4 DNA ligase. T4 DNA ligase can join the broken blunt ends of two DNA molecules irrespective of the sequences. The disadvantage of this technique is that any two broken ends may join including those belonging to same DNA molecule. However, blunt-ended ligation is a useful way of joining together DNA fragments, which have not been produced by the same restriction enzyme and therefore probably have incompatible sticky ends. These ends are removed prior to ligation using the enzyme S1 nuclease, which digests single stranded DNA.

Blunt ligation is most usefully applied where blunt-ended fragments are joined via *linker* molecules (described later). The linker molecule can be ligated to both ends of foreign DNA to be cloned, and then treated with restriction enzyme to produce sticky ended fragments. The vector molecule is also cut with same restriction enzyme. Finally, the foreign DNA is ligated to vector via linker created sticky ends. These linkers also help in excising and recovering the foreign DNA (gene of interest) after cloning and amplification in host bacterium. Here a brief account on various types of linkers is given.

Linkers

They are self complementary synthetic oligonucleotides which contain restriction endonuclease recognition sites. They may be used to clone DNA fragment into plasmids or other vectors.

These blunt ended linkers are first phosphorylated using polynucleotide kinase and labelled or unlabelled ATP molecules. These are then ligated to the blunt ended DNA fragment using T4 DNA ligase. The DNA fragment is finally treated with restriction enzyme to generate sticky ends and ligated into vector which has been cut by the same enzyme.

Stop Linker

They are 5'-phosphorylated double-stranded synthetic oligo-deoxynucleotides which can be cloned into an expression vector to ensure transcription termination. Such linkers contain stop codons in each reading frame from both directions. Each stop linker contains an SspI restriction site, which can be used to screen clones for linker insertion.

Adaptors

They are linkers which contain more than one restriction enzyme recognition sites. They are used to insert new restriction sites into cloning vectors or to join restriction fragments to vectors with incompatible cohesive ends. (Fig. 4.2).

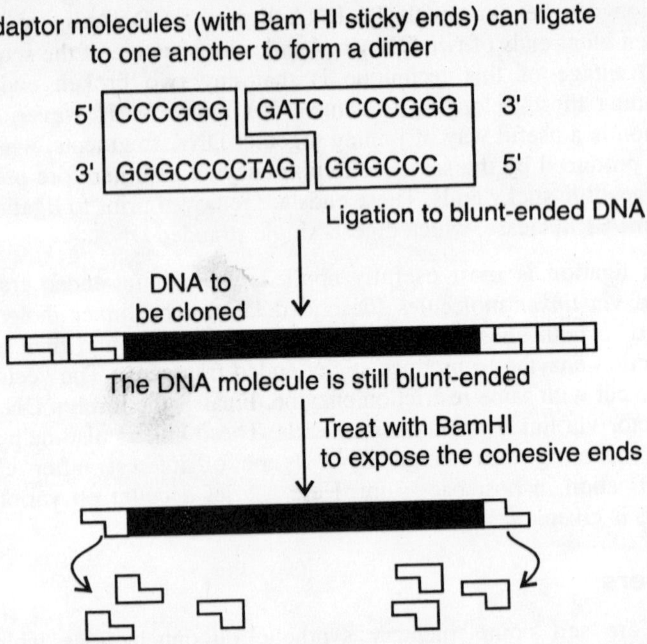

Fig. 4.2: Digram to explain the problems associated with the use of adoptor molecules. A. Two adoptor molecules with Bam HI cohesive ends have ligated to each other to form a dimer. B. Such dimer of the adoptor molecules can ligate to a blunt-ended DNA molecule. The DNA molecule thus obtained is still blunt-ended. C. The cohesive ends are recreated by treatment with Bam HI.

Double-linkers

In some plasmid vectors like pUC8, duplex cDNA molecules can be cloned. This is done by double-linker approach, in which different linker molecules are added to the opposite ends of the cDNA. This is an advantageous method as:

1. Orientation of inserted DNA is fixed.
2. Vector reclosure without insertion of foreign DNA can be done.
3. Efficient for cloning of cDNA obtained from rare mRNA molecules.
4. Better than homopolymers where expression from vector-borne promoter is desirable.

Terminal Transferase

Besides adding cohesive ends or linkers, blunt ended DNA fragment and vector can also be ligated. This technique makes use of producing and annealing complementary homopolymer sequences. An enzyme, terminal transferase (purified from calf-thymus), is used to synthesize homopolymeric extensions. Using precursor dGTP, poly dG is added at both the 3'cut ends of vectors while poly dC is added to 5'cut end of foreign DNA. The vector and clone can then be joined by annealing the poly dG with poly dC tails and then ligating them using DNA ligase enzyme. The cloned sequence can be retrieved by using Pst I ŕestriction enzyme. The method works out the chances of reannealing between the two cut ends of same DNA molecule or of two similar DNA molecules, thus is advantageous.

CLONING IN BACTERIA AND EUKARYOTES

Cloning in Bacteria

After ligation of target into vector, the recombinant DNA is multiplied in a host. Various bacterial strains are available as host. Most common prokaryotic host is *E. coli*, which is also termed as **workhouse of genetic engineering**. In this technique, the recombinant DNA is subjected to *in vitro* packaging and used to transform bacteria for its amplification. The host bacteria are modified and selected on the basis of high transformation efficiency. The type of bacterial host is selected on the basis of plasmid vector. Various bacterial hosts commonly employed are :

1. **Gram positive bacteria e.g. *E. coli***: host for plasmids only.
2. **Gram negative bacteria e.g. *Bacillus subtilis*:** host for plasmid, phages and cosmid vectors.

Chimeric DNA is inserted into bacterial cells to create transformed bacterial cells. Colonies of transformed cells are selected and used for multiplication in suspension cultures. The recombinant DNA is retrieved and isolated from the host cell when required.

Cloning in Eukaryotes

E. coli is workhouse of genetic engineering but, the chromosome structure and gene expression in prokaryotes and eukaryotes differ radically. Therefore, *E. coli* offers limitation to its use in cloning eukaryotic genes. The split genes can not be correctly expressed in prokaryotic host as latter do not have introns and thus splicing of mRNA can not take place. Therefore, eukaryotic cells are used to clone and express eukaryotic genes.

Among eukaryotes, DNA cloning has been done in yeast, mouse and some plant species. Yeast has following attractions as host for genetic engineering:

1. Short doubling time (90 minutes).
2. Can be maintained in either a haploid or diploid state.
3. Can be grown on simple defined media and auxotrophs can be selected.
4. The wall can be removed and foreign DNA entry can be facilitated with PEG and $CaCl_2$ mixture.
5. Confers extra-chromosomal replication of chimeric DNA just as plasmids in *E. coli*.

The simplest kind of yeast vector contains a selectable marker such as TRP1, joined to plasmid vector. TRP1 encodes an enzyme in the tryptophan biosynthetic pathway and when introduced into TRP^- cell by DNA transformation, it renders the cell capable of growth in absence of added tryptophan. Vector used is a hybrid or shuttle type. When transformed into yeast, such a vector integrates into the yeast genome by recombination and makes the transformed yeast selectable on the basis of growth in absence of added tryptophan.

Yeast is becoming an important organism for the commercial production of medically important proteins such as viral vaccines. It is of great value as an organism for basic research, for producing heterologous proteins and as a vehicle for cloning large segments of DNA. For some purpose, however, it is essential to be able to transform mammalian cells. While yeast and mammals share many basic properties, these are important different cells in the precise mechanics of gene expression. Mammalian RNA transcripts are not spliced in the yeast cell, nor the mammalian proteins are glycosylated correctly.

In animal cells like mouse cells, special animal viruses are used as cloning vehicles. Simian virus 40 (SV40) with foreign DNA can be transcribed and translated in mouse kidney. Retroviral vectors are used for transformation of mammalian cells. These animal viruses comprise single stranded RNA genome, gene of interest and G418R gene for resistance against protein synthesis inhibitor G418.

AMPLIFICATION OF DNAs BY POLYMERASE CHAIN REACTION (PCR)

The major scientific development of year 1989, as named by prestigious journal 'Science' was **Polymerase Chain Reaction** (PCR). Kary Mullin is the inventor of PCR, who works in Cetus Corporation, a biotech company, of United States. Since its introduction, the polymerase chain reaction has revolutionised the method of DNA analysis in both research and clinical laboratories.

The PCR, is an extremely powerful new procedure that allows one to make million fold copies of a selected DNA sequence in a genome. PCR can be used to clone a given DNA sequence *in vitro* without using living cells during the cloning process. However, the procedure can only be applied when the nucleotide sequence of at least one short DNA segment on each side of the 'region of interest' is known. The PCR procedure makes use of synthetic oligonuclotides complementary to the known sequences spanning the region of interest to fetch *enzymatic amplification of this segment of DNA in the test tube.*

PCR may eventually replace gene cloning technique. This technology is much easier and requires much smaller quantity of DNA (in ng) whereas for cloning, DNA is needed in microgram (μg) quantities.

Procedure

The PCR procedure involves three steps, each repeated many times to produce cycles of amplification.

Step 1: The genomic DNA containing the sequence to be amplified is denatured by heating.

Step 2: The denatured DNA is annealed to an excess of the synthetic oligonucleotide primers.

Step 3: DNA polymerase is used to replicate the DNA segment between the sites complementary to the oligonucleotide primers. The primer provides the free 3'-OH required for covalent extension; the denatured genomic DNA provides the required template function. These three steps (denaturation, primer binding and DNA synthesis) represents a single PCR cycle.

The products of the first cycle are replicated for further amplification. This reaction can be performed many times to supply unlimited copies of amplified DNA. Repetition is made possible by regular denaturation of freshly synthesized double stranded DNA molecules by heating it to 90-98°C.

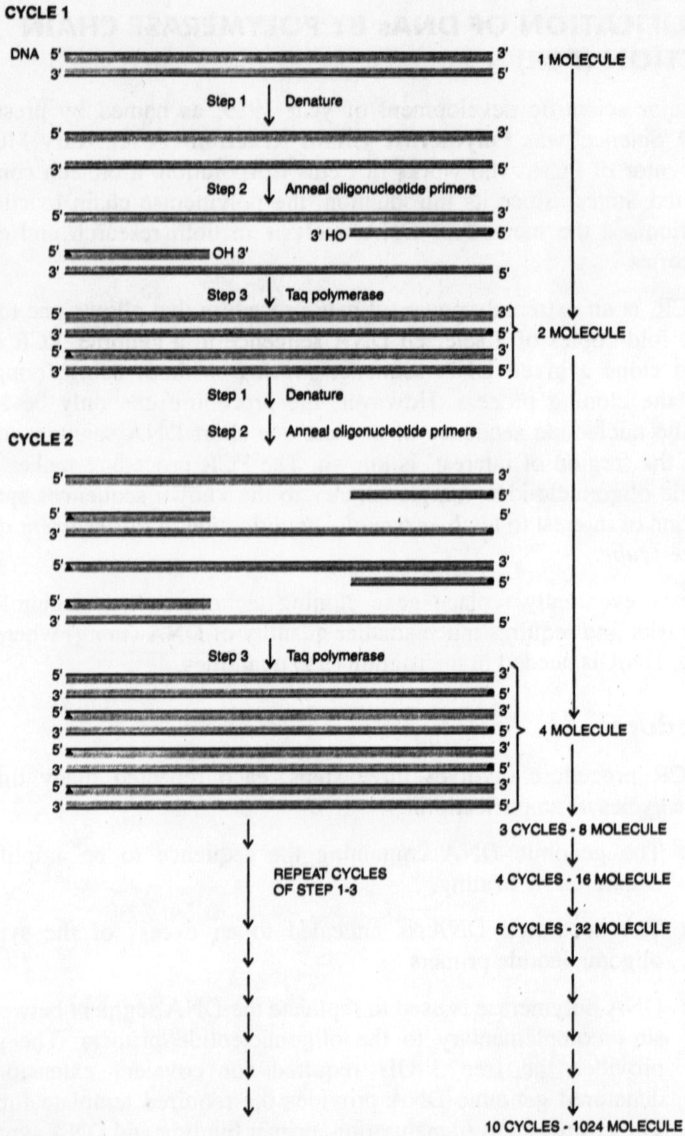

Fig. 4.3: Diagram illustrating the use of PCR to amplify DNA molecules *in vitro*. Each cycle of amplification involves three steps: (1) denaturation of the genomic DNA being analyzed, (2) annealing of the denatured DNA to chemically synthesized oligonucleotide primers with sequences complementary to sites on opposite sides of the DNA region of interest, and (3) enzymatic replication of the region of interest by Taq polymerase.

At this temperature, the two strands separate. Once the double stranded DNA is made single stranded, it is mixed with two primers and is cooled to 40-60°C.

This allows the primers (which are in excess) to bind to their complementary strands through renaturation followed by replication with DNA polymerase. The cycle is repeated many times until the desired level of amplification is achieved.

Applications of PCR

During the last few years, with the improvement of PCR protocols and due to the availability of automatic thermal cyclers commercially, the applications of PCR have increased manifold. Some of the applications are listed here.

Study of DNA polymorphism using PCR

DNA polymorphism is amplified using PCR and can be used for study. Loci either with known DNA sequence or at random sites, both can be amplified. PCR can be used to study RFLPs as well as RAPDs.

Loci with poly (A), poly (G), poly (TC) and poly (TG) etc., known as short tandem repeats or simple sequence repeats (SSR) can also be amplified with PCR. Some of the these SSR loci such (CA), or (GT)n occur in the human genome with many copies. SSRs can be utilized in preparing saturated molecular maps.

Gene tagging using PCR

PCR has also been utilized for developing molecular markers closely linked to specific genes of economic importance. For instance in tomato, 144 random primers were used to produce 625 PCR products from a set of near isogenic lines. Near isogenic lines (= NIL's) differ only for the presence and absence of a specific gene for *Pseudomonas* resistance. Atleast three of these PCR products were present in one and not in the other of a pair of NIL's showing linkage with *Pseudomonas* resistance gene (Pto). These PCR products could be mapped close to Pto.

PCR for confirming the presence of transferred gene

When a chimeric gene is transferred to cultured cells or organisms, primers can be designed to conduct PCR for amplification of the gene sequence so transferred.

This technique allows the confirmation of transfer and maintenance of the gene of interest.

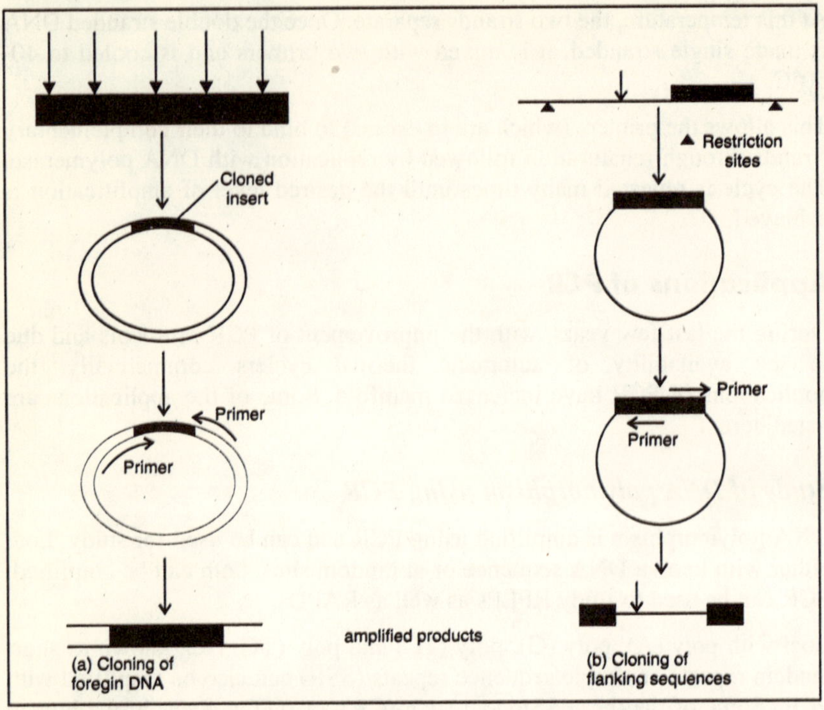

Fig. 4.4: Inverse PCR (a) amplification of genes using flanking vector sequences as primers; (b) amplification of sequences flanking the gene, using sequences at the two ends of the gene as primers.

DNA fingerprinting using PCR

DNA fingerprinting is used in forensic medicine for identification of criminals and disputed parentage. Some well characterized sequences of microsatellites are being used for designing primers. PCR allows amplification of DNA from individual hairs, stains of blood or seminal fluid having partially degraded DNA. Characterization of individuals with such small amount of DNA could not be done earlier.

PCR in Gene Therapy

In gene therapy experiments, the transfer and presence of a marker gene for neomycin resistance (NeoR) could be detected in the blood of patients, even after 60 days, although the frequency of marked lymphocytes gradually decreased necessitating a second transfusion of such marked cells. Similarly, ADA (adenosine deaminase deficiency) gene was transferred for therapeutic purposes and PCR was used to detect the presence of this gene. Thus, PCR proves to be of immense help in monitoring a gene in gene therapy experiments.

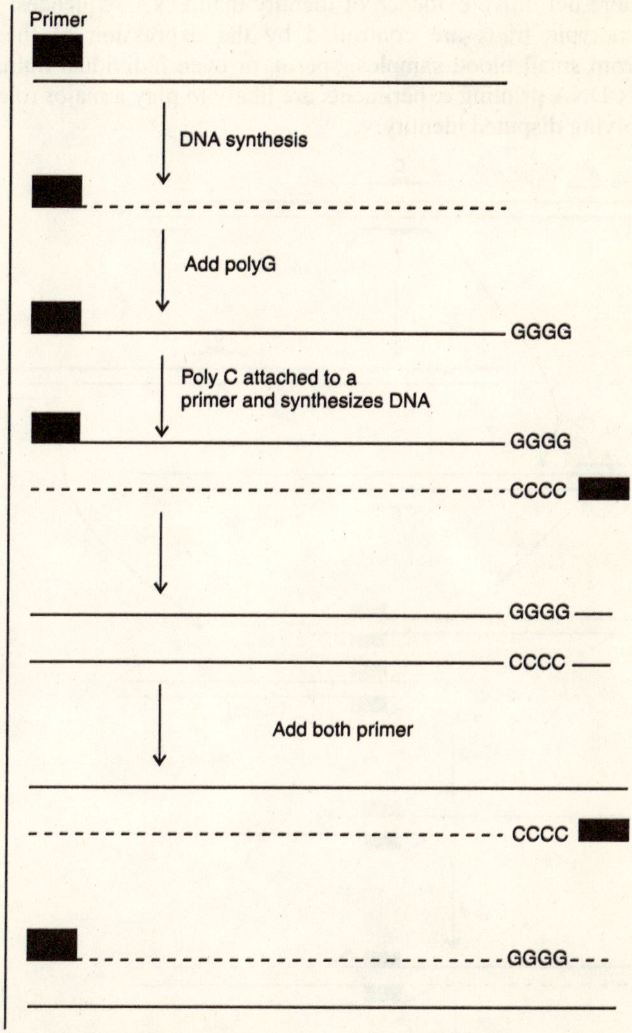

Fig. 4.5: Anchored PCR, using a single primer in the first cycle.

PCR technologies thus provide shortcuts for many cloning and sequencing applications. These procedures permit scientists to obtain definitive structural data on genes and DNA sequences when very small amounts of DNA are available. One important application involves the diagnosis of inherited human disease, especially in cases of prenatal diagnosis, where very limited amounts of fetal DNA are available. A second major application occurs in forensic cases involving individual identities. No criteria can provide more definitive evidence of identity than DNA sequences; after all, other phenotypic traits are controlled by the expression of these DNA isolated from small blood samples, sperm, or even individual human hairs. Thus, PCR DNA printing experiments are likely to play a major role in legal cases involving disputed identity.

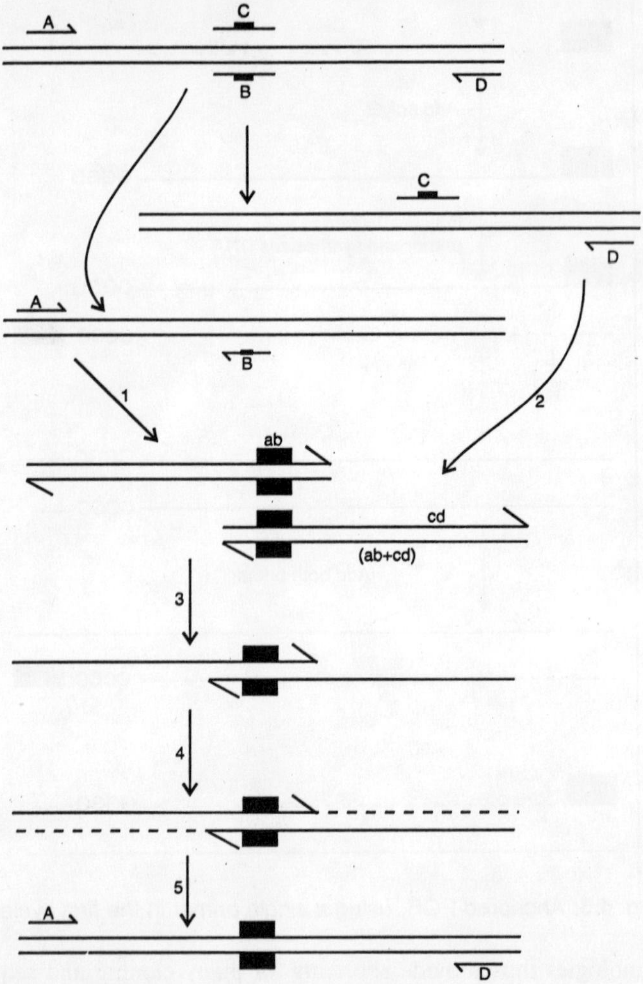

Fig. 4.6: Introduction of mutation at an internal site using overlap extension (two pairs of primers A & B and C & D are used. B and C represent internal sites and overlap each other).

GENE TRANSFER TECHNOLOGY

Gene transfer is of basic as well as applied interest. It enables to study the structure and function of isolated genes via modification, reintroduction and subsequent analysis of their function in transgenic plants, cell lines or even cell populations. Besides these, transfer of specifically designed hybrid genes to plant genomes creates novel varieties with designed characters.

Gene transfer in plants can be accomplished by sexual and asexual methods. Asexual methods include:

1. Non-specific gene transfer by protoplast techniques.
2. Specific gene transfer methods.

In animals, gene transfer is done using egg and cultured stem cells or by specific gene transfer.

Sexual Method

Gene transfer using pollen or pollen tube

Pollen grains are treated with heavy dose of X-ray which pulverizes chromosomes. The pollen, if able to pollinate, assures incorporation of changed chromosomes into daughter progeny. It is assumed that DNA can be taken up by the nuclei or reach the zygote through the pollen tube pathway. Both these approaches have been tried, though several problems exist in this method like presence of cell wall, nucleases, callose plugs in pollen tube etc.

Incubation of dry seeds, embryos, tissues or cells in DNA

This method has been tried in many cases and expression of defined genes has been witnessed. In all the cases, plant cell walls work as efficient barriers, thus DNA uptake can be enhanced by PEG or electroporation or by any other device.

Asexual Methods

Non Specific Gene transfer

Fusion of protoplasts to form somatic hybrid and partial hybrids are the known methods. The most direct method of gene transfer to plant cells is the uptake of plasmid DNA but at present its application relies on the availability of isolated genes. Agronomic traits are not coded for by single genes but rather by families of interacting genes. Thus, there is an interest in techniques which allow the transfer of large blocks of intact DNA encoding polygenic traits.

Protoplast membranes may be induced to fuse with nucleated or mitotically inactivated protoplasts to form hybrids; they may be induced to take up isolated nuclei or chromosomes; or they may be fused with bacterial spaheroplasts, or liposomes.

Protoplast fusion may be carried out with three different methods; by using chemicals, by an electric field or by using a U.V. laser microbeam. Electrofusion is reported to be free of cytotoxicity, associated with chemical fusagen such as polyethylene glycol (PEG) or polyvinyl alcohol (PVA). Electrofusion appears superior to laser fusion as former is commercially available and affordable for a typical biological laboratory while this is not presently true for the laser microbeam. Furthermore, the number of hybrids that can be produced in a given time is much higher using electrofusion.

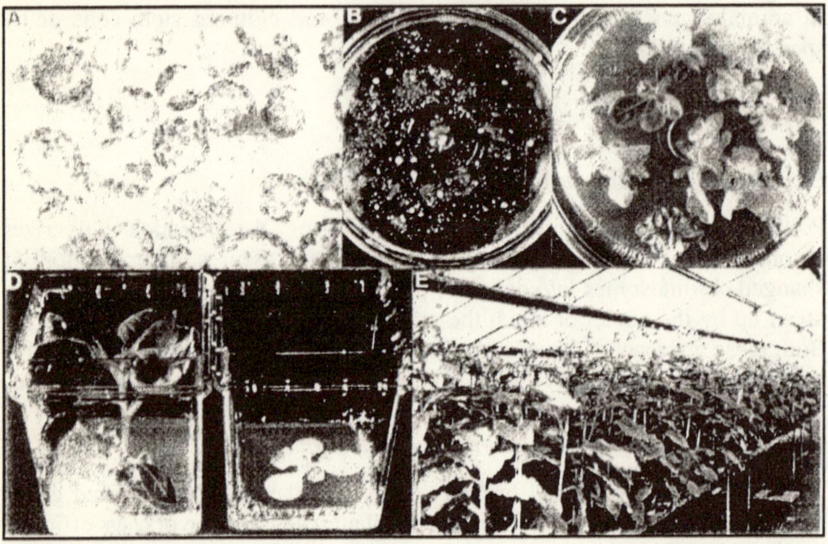

Fig. 4.7: Direct transfer to leaf protoplasts and regeneration of transgenic plants in tobacco (*Nicotiana tabacum*). (A) Freshely isolated leaf protoplasts of tobacco treated with polyethylene glycol and plasmid DNA bearing a chimeric *npt2* gene as selectable marker; (B) selection of kanamycin-resistant transformed colonies from protoplasts in bead-type culture; (C) shoot differentiation from tobacco calli obtained from transformed protoplasts; (D) rooting of transformed (left) and untransformed (right) shoot from protoplasts on selective medium; (E) fertile transgenic tobacco plants obtained from protoplasts.

Foreign DNA can be directly introduced into mammalian cells. Firstly, DNA is precipitated with $CaCl_2$ and then this precipitate is mixed with the cells to be transformed. The DNA will be taken up by the cells. Inside the cell, the new DNA fragments get ligated to give a concatamer, which is then integrated into the nuclear DNA as one single piece. The integration occurs at random sites. In order to select the transformed cells several marker genes are also introduced along with the genes to be cloned. The marker genes will be linked with the genes of interest prior to integration into the host genome. One such marker is the neomycin resistance gene, which would normally inhibit protein synthesis.

Specific Gene Transfer

Specific gene transfer in plants is mostly done with *Agrobacterium* species having natural biological delivery system. There are several other techniques developed recently for introducing cloned genes into plant cells. These are categorized as artificial delivery system transferring gene directly. Some examples of artificial delivery systems are direct DNA uptake, electroporation, fusion with liposomes etc. Unlike with *Agrobacterium* species mediated gene transfer (where cells with intact cell walls and tissue explants can be used), artificial delivery systems can be applied to protoplasts only.

The basic pre-requisites for successful gene transfer are:

1. selectable marker gene, which could be expected to yield a clear phenotype in transformed cell clones;
2. expression signals, which are independent of physiological state of host cell;
3. precise linkage of the protein-coding region of the selectable gene to the expression signals. This will guarantee the expression of hybrid gene in the host genome;
4. DNA stretches on both sides of the hybrid gene for its integration into the host genome without disrupting the gene construct;
5. an absolutely tight selection protocol for presumptive transformants; and
6. where of artificial delivery systems are utilized, the protoplast system taken should have routine high plating efficiency and plant regeneration capacity.

Different methods for specific gene transfer are discussed in this chapter.

Biological Delivery System

To date, *Agrobacterium. tumefaciens* mediated transformations have been in wider practice and a much broader range of transgenic plants have been produced with this technique. It is a method of choice for routine production of transgenic species since it can transform regenerable tissues with relative ease.

Agrobacterium tumefaciens: An Important Vector for Genetic Engineering

Agrobacterium tumefaciens is the most important tool in the genetic engineering of plants (the Ti plasmid of the soil bacterium *A. tumefaciens*). This bacterium is the causative agent of crown gall disease of dicotyledonous plants only (unable to develop gall in monocotyledonous species).

The name refers to the galls or tumours that often form at the crown (junction between the root and the stem) of infected plants. Since the crown of the plant is usually located at the soil surface, this is where a plant is most likely to be wounded (e.g., due to a soil abrasion from the plant blowing in a strong wind) and infected by a soil blowing in a strong wind) and infected by a soil bacterium such as *A. tumefaciens.* However, *A. tumefaciens* can infect a plant and induce a tumour at any wound site.

In the normal disease process the bacterium enters the plant at the site of a wound. The infection often occurs at the crown, but it may involve an upper part of the stem, the petiole, or the leaf. The bacterium attaches to the wall of a cell and transforms the cell. Infected cells proliferate at the site of infection, forming a tumor or gall. The tumor can be fatal to young plants if it grows big enough to crush the vascular system, although often the tumor is merely debilitating and unsightly. Transformed cells also have the ability to grow in culture on a medium that lacks the hormones auxin and cytokinin. Finally, transformed cells produce opines, such as octopine and nopaline. These are chemicals that are formed from two amino acids. Argenine and alanine form octopine, and argenine and glutamine produce nopaline. Octopine and nopaline are not found in normal plant cells.

Infection of Agrobacterium tumefaciens

During the infection of a wound site by *A. tumefaciens,* following two key events occur:

1. The plant cells begin to proliferate and form tumours.
2. They begin to synthesize an *arginine* derivative called an *opine.*

The synthesized opine is usually either nopaline or octopine depending on the *A. tumefaciens* strain involved.

These opines are catabolized and used as energy sources by the infecting bacteria. Interestingly, *A. tumefaciens* strains that induce the synthesis of nopaline can grow on nopaline, but not on octopine, and vice versa. Clearly, an interesting interrelationship has evolved between *A. tumefaciens* strains and their plant hosts. *Agrobacterium tumefaciens* is able to divert the metabolic resources of the host plant to the synthesis of opines, which are of no apparent benefit to the plant, but which provide sustenance to the bacterium.

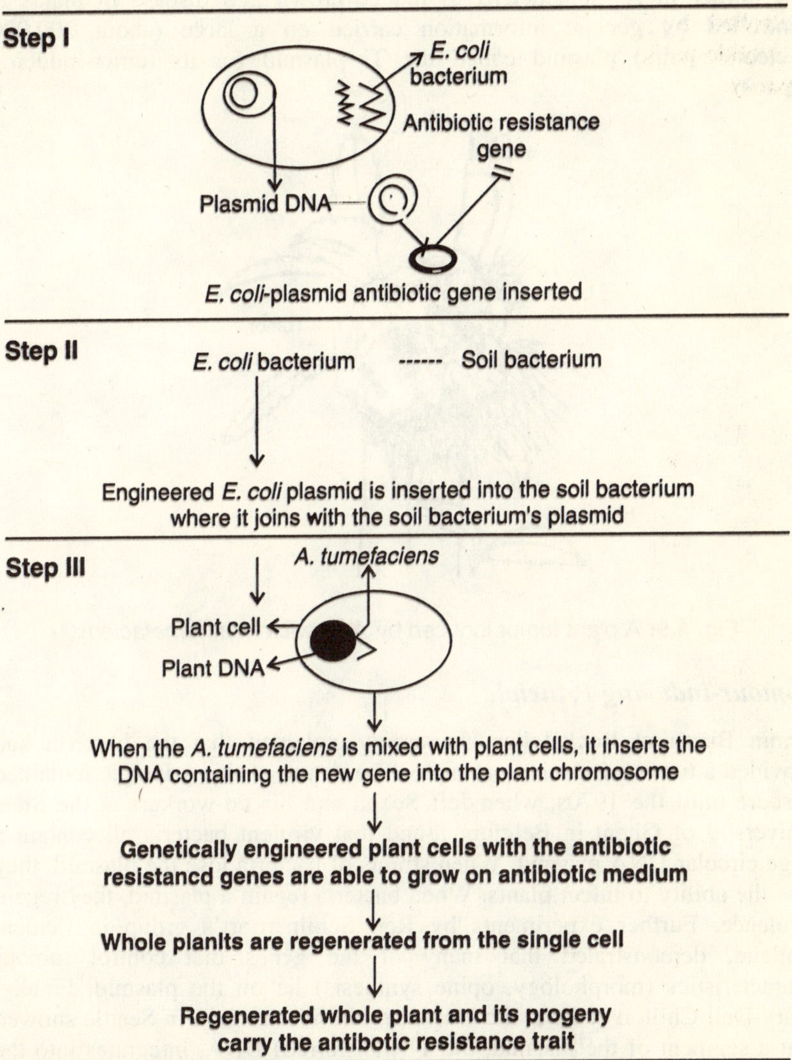

Fig. 4.8: Various steps (I-III) used in plant genetic engineering system.

Genetic Information Needed for Crown Gall Development (Ti Plasmid)

The ability of *A. tumefaciens* to induce crown gall disease in plants is controlled by genetic information carried on a large (about 200,000 nucleotide-pairs) plasmid called the Ti plasmid for its tumor-inducing capacity.

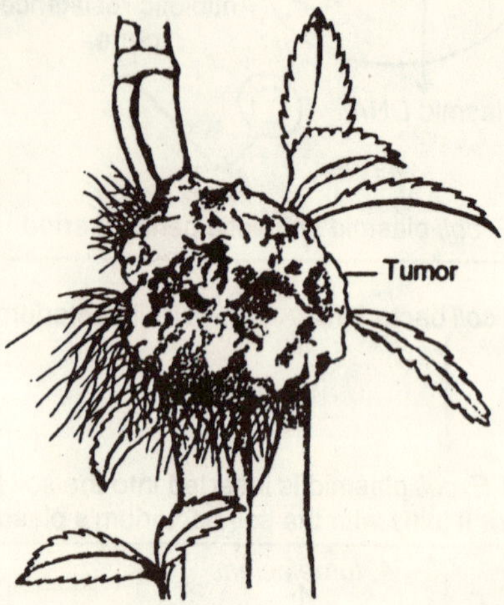

Fig. 4.9: A plant tumor induced by *Agrobacterium tumefaciens.*

Tumour-inducing Principle

Armin Braun at Rockefeller University explained that the bacteria had provided a tumour-inducing principle. The nature of this principle remained obscure until the 1970s, when Jeff Schell and his co-workers at the State University of Ghent in Belgium found that virulent bacteria all contain a large circular DNA plasmid. When strains of bacteria lose the plasmid, they lose the ability to infect plants. When bacteria regain a plasmid, they regain virulence. Further experiments by Rob Schilperoort's group in Leiden, Holland, demonstrated that many of the genes that control tumour characteristics (morphology, opine synthesis) lie on the plasmid. Finally, Mary-Dell Chilton et al. at the University of Washington in Seattle showed that a segment of the plasmid, the T (transferred) DNA, integrates into the genome of transformed cells. The plasmid (Ti plasmid) is thus identified as the tumour-inducing principle.

Two components, the T-DNA and the vir region, of the Ti plasmid are essential for the transformation of plant cells.

T-DNA (Transferred DNA)

During the transformation process, the T-DNA (for Transferred DNA) is excised from the Ti plasmid, transferred to a plant cell, and integrated (covalently inserted) into the DNA of the plant cell. The available data indicate that integration of the T-DNA occurs at random chromosomal sites; moreover, in some cases, multiple T-DNA integration events occur in the same cell. In nopaline-type Ti plasmids, the T-DNA is a 23,000-nucleotide-pair segment that carries 13 known genes. In octopine-type Ti plasmids, there are two separate T-DNA segments.

Some of the genes on the T-DNA segment of the Ti plasmid encode enzymes that catalyze the synthesis of phytohormones (the auxin-IAA and the cytokinin isopentenyl adenosine). These phytohormones are responsible for the tumorous growth of cells in crown galls. The T-DNA region is bordered by 25-nucleotide pair imperfect repeats, which are required in cis for T-DNA excision and transfer. The deletion of either border sequence completely blocks the transfer of T-DNA to plant cells.

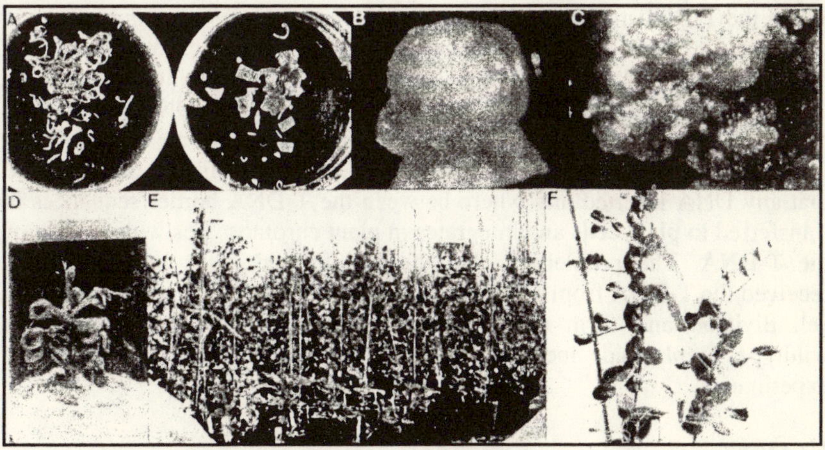

Fig. 4.10: *Agrobacterium*-mediated gene transfer and regeneration of transgenic plants in pea (*Pisum sativum*). (A) co-cultivation of leaf and shoot explants of pea with *A. tumefaciens* strain bearing Ti-plasmid with chimeric *hpt* gene as stable marker; (B) Selection of transformed pea callus on hygromycin-containing proliferation medium; (C) induction of shoot formation from transformed pea callus; (D) in vitro growing transformed pea plant; (E) adult transgenic pea plants growing in soil under greenhouse conditions; (F) fertile transgenic pea plants obtained from *Agrobacterium*-mediated gene transfer.

Vir *Region (For Virulence)*

The *vir* (for virulence) region of the Ti plasmid contains the genes required for the T-DNA transfer process. These genes encode the DNA processing enzymes required for excision, transfer, and integration of the T-DNA segment. *vir* genes can supply the functions needed for T-DNA transfer when located either *cis* or *trans* to the T-DNA. The *vir* genes are normally expressed at very low levels in *A. tumefaciens* cells growing in soil. However, exposure of the bacteria to plant cells or exudates from plant cells induces increased levels of expression of the operons containing the vir genes. Surprisingly, this induction process is very slow for bacteria, taking 10 to 15 hours to reach maximum levels of expression. The phenolic compounds like acetosyringone have been purified from plant exudates and shown to act as inducers of the *vir* operons. Thus, the induction process can now be studied in vitro by using these purified inducer molecules.

Disarmed Ti Vectors

When the T-DNA region of the Ti plasmid of *Agrobacterium tumefaciens* is transferred to plant cells and becomes integrated in plant chromosomes, the potential use of *A. tumefaciens* in plant genetic engineering seems to be obvious. The foreign genes can be introduced into the T-DNA region and, hopefully, these genes can be transferred to the plant with the rest of the T-DNA segment. In fact, this works very well. All the available data indicate that any DNA inserted any where between the T-DNA border sequences is transferred to plant cells and integrates in plant chromosomes with the rest of the T-DNA. The problem is that the transformed plant cells that have received the T-DNA from a wild-type Ti plasmid lose their normal control of cell division and form tumours. This oncogenicity of T-DNA renders wildtype Ti plasmids incompatible with the goals of most gene-transfer experiments.

Vectors

A shuttle vector is commonly used, which can replicate in *E. coli* (where recombinant manipulations can be easily handled), and then transferred into *A. tumefaciens* in preparation for transfer into plants. There are two basic modes of maintenance of the shuttle vectors in *Agrobacterium*:

1. either by integration into Ti plasmid by recombination at a region of DNA homology
2. or, by autonomous replication in trans to the Ti plasmid.

The former type of vector is referred to as cis or integrating vector while the latter is called a trans or binary vector.

A selectable markers is required to recognize the transformed cells and on other hand, suppress the growth of wild type cells. This helps in identification of transformed cells, which are in small proportion.

Limitations: *A. tumefaciens* based gene transfer is hampered by host range limitations, which are especially disturbing when aiming at gene transfer into cereals or other graminaceous monocots.

Artificial Delivery Systems

Escape from host range limitations can be made by transferring the genes *directly* without using any biological vectors, into plant cells. Cell wall constitutes an effective barrier for DNA molecules to pass; naked plant cells, the 'protoplasts', have been found ideal experimental system to study vectorless gene transfer. Various artificial delivery systems are given here.

GENE TRANSFER USING PARTICLE BOMBARDMENT TECHNIQUE

The first genetically engineered plants were produced in middle 1980s (Horsch et al. 1985) and developed interest of genetic engineers in the production of transgenic plants for both basic and applied work. The major efforts were in the development and refinement of methods for transgenic plant production and also in the isolation and characterization of useful gene for introduction into plants. Through gene transfer, plants have already been produced that contain genes for disease and insect resistance, and modified fruit and grain quality. Particle bombardment genes transfer (or biolistics) is a method of transferring genes. Biolistics has been used to transform a large number of different plant species (reviewed by Christou 1994), as well as some animals, fungi, and bacteria.

The basic principle of particle bombardment for gene transfer is to use high velocity microprojectile to incorporate the introduced genes. The outer cell layers are penetrated in order to introduce genetic material into living cells.

History

Biolistics (a combination of *biological* and *ballistics*), is a method in which high-velocity microprojectiles are used to deliver exogenous genetic material into plant tissue. It was developed, and first described by John Sanford in collaboration with Ted Klein, Ed Wolf, and Nelson Allen (Sanford et al. 1987). The first used penetration particles was onion tissue, where a monolayer of large epidermal cells was used as the target tissue. By observing cytoplasmic streaming in these cells, this tissue could be easily evaluated for viability following bombardment. Onion epidermal cells, transformed with the chloramphenicol acetyl transferase (CAT) gene were the first transiently transformed cells produced using this method.

The first stably transformed plants obtained using biolistics were reported by Christou et al. (1980) and Klein et al. (1988) using soybean and tobacco respectively. Since these initial reports of successful transformation, a broad range of species has been transformed including bacteria (Smith et al. 1992), fungi (Toffaletti et al. 1993) and even animals et al. 1988). In addition, organelles such as chloroplasts (Boynton et al. 1988) and mitochondria (Johnston et al. 1988) have also been transformed using Biolistics.

This method of particle acceleration, has been described as the *particle gun, the gene gun, the bioblaster*, and the *microprojectile bombardment method.* It initially christened the biolistic method by its inventor (Sanford, 1988). Biolistics is a technique which utilizes instrumentation to accelerate DNA coated microprojectiles into cells, past the cell wall and cell membrane. The microprojectile is small enough (0.5-5μm) to enter the plant cell without too much damage, yet large enough to have the mass to penetrate the cell wall and carry an appropriate amount of DNA on its surface into the interior of the plant cell. The important aspects of biolistics are:

1. It is a direct gene transfer method.
2. It does not require biological vector.
3. It has been used successfully to transform cells that were once impossible to transform by other means.
4. DNA delivery appears to be genotype and tissue independent.
5. With this technique, the microprojectile is shot through the physical barriers of the cell.
6. There are no biological limitations to DNA delivery.
7. Transformation via the biolistics method is simple and safe.

Applications of Biolistic

1. The utility of the Biolistic method is broad and the technology can be used for basic studies on transformation.
2. Modifications of gene expression.
3. Understanding essential components of genes.
4. The production of transgenic plants of economic value.

Particle Gun Design

Several particle gun have been designed in various laboratories. The basis of all of these designs is to coat DNA onto small dense particles and accelerate the particles towards a target tissue. The particles usually consist of either gold or tungsten.

Gold particles are chemically inert. They do not produce cytotoxic effects. They are more uniform in size than tungsten particles. However, the cost of gold particles may limit their accessibility and the more affordable tungsten particles are sufficient for most studies.

Tungsten particles are somewhat phytotxic (RUSSELL et al. 1992a) and tend to be more variable in size than gold particles. Ideally the particles used for bombardment should have good initial affinity for the DNA, yet freely release it once in the cytoplasm or nucleus of the target cell.

Preparation of DNA-coated Microprojectiles

1. Wash gold or tungsten particles and mixed them with plasmid DNA.
2. The DNA is bound on the particles using either ethanol or $CaCl_2$ precipitation methods.
3. Spermidine may be added to the mixture, possibly protecting the DNA from degradation and/or altering its conformation.
4. After precipitation, the particles may be washed, resuspended and either dried or stored on ice as an aqueous suspension until use.

Use of Macrocarrier

Many of the particle guns utilize a macrocarrier. They supports or carries the particles and gun is accelerated along with the particles towards the target. The macrocarrier is usually retained by a stopping plate or screen before it collides with the target, whereas the particles continue along their course. In most cases, the particles are accelerated under partial vacuum in a vacuum chamber to reduce air drag.

The penetration of particle is controlled by modifying the intensity of the explosive burst. It is done by changing the distance that the particles must travel to reach the target tissue or by using different sized particles.

Gene Gun

The original gene gun described in the literature used DNA-coated tungsten particles in an aqueous slurry loaded on the end of a small plastic bullet, which served as the macrocarrier. The plastic macrocarrier was placed into a 0.22 caliber barrel in front of a gun powder cartridge. Once the cartridge was fired, the plastic bullet was propelled towards a solid stopping plate containing a small opening in the center. The microprojectiles continued through the hole towards the target tissue while the macrocarrier was retained. Both the stopping plate and the macrocarrier were discarded afterwards. The initial blast from the gunpowder cartridge necessitated frequent cleaning of the chamber and the target variation in results were common with this device, this first unit was used successfully by a number of different laboratories.

Modified Bombardment Devices

Helium-Modified Bombardment Device

The first important modification of the original device included the substitution of a helium blast for the gunpowder discharge (Sanford et al., 1991). Helium pressure continually builds in a reservoir and is released using *rupture discs.* The release of helium produces a shockwave which travels to a second disc or macrocarrier. It holds the DNA-coated particles. The macrocarrier carries the DNA coated particles a short distance into a stopping screen. It retains the macrocarrier. The microprojectiles continue traveling, ultimately penetrating the target tissue which is held in a partial vacuum. The PDS-1000/He is the most commonly used device.

Accel Particle Gun

This gun uses a high voltage electrical discharge to vaporize a water droplet which produces a controlled shock wave (Mccabe and Christou, 1993). The device is constructed so that the initial shock wave is reflected to produce secondary shock waves. This in turn accelerate a mylar sheet coated with particles. The mylar sheet is accelerated towards a retaining screen, stops the carrier and allows the particles to continue onto the target tissue. Particle speed, and the resulting depth of particle penetration are accurately modified by adjusting the intensity of the voltage passing through the water droplet. It is the most efficient particle bombardment device in current. However, it is not commercially available.

Particle Inflow Gun

It is a costly device due to manufacturing expenses. This created a need to develop more affordable, simple, yet safe and effective devices such as the **Particle Inflow Gun** or PIG (Finer, et al., 1992). With this device, a solenoid controlled by a timer relay is used to generate a burst of low-pressure helium. The particles, supported by a screen in a reusable syringe filter unit, are accelerated directly in a stream of helium without the need for a macrocarrier. The target tissue is held in a chamber under a partial vacuum.

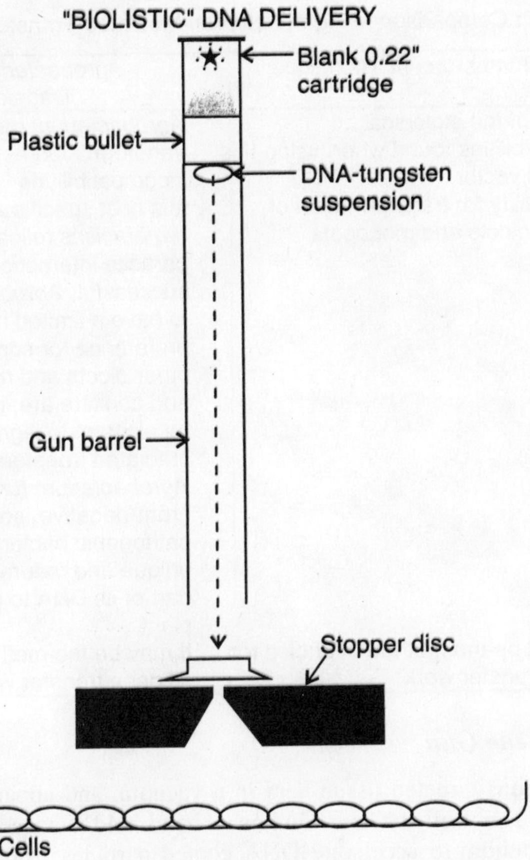

Fig. 4.11: The gene gun.

To minimize the localized impact of the particles and to aid in dispersal, a nylon mesh baffle may be placed over the tissue. For this device there is no need to purchase macrocarriers or rupture membranes. It is less cleanup and minimal down time between shots.

Microtargeting Device

This device accelerates small amounts of a DNA/particle mixture in a focused stream of high-pressure nitrogen. It was originally developed to allow precise delivery of particles to the shoot meristem. DNA is not precipitated on the gold particles but is used as a DNA/particle mixture. If the shoot meristem could rapidly give transgenic progeny.

Table 4.1: Comparison to *Agrobacterium*-Mediated Transformation

Particle Bombardment Technique	*Agrobacterium*-Mediated Transformation
No need of the biological incompatibilities found when using this biological vector	*Agrobacterium* medicated transformation needs biological incompatibilities
A good utility for transformation of conifers, dicots and monocots	It is host specific *Agrobacterium tumefaciens* relies on a host-parasite interaction in order to be successful. *Agrobacterium* appears to have a limited host range with a preference for some dicots while other dicots and most monocots and conifers are, in general more recalcitrant to *Agrobacterium*-mediated transformation. *Agrobacterium tumefaciens* is a gram-negative, soil-borne pathogenic bacterium that has the unique and natural ability to transfer part of its DNA to the cells of a host plant
It may not be the method of choice for all gene transfer work	It may be the method of choice for all gene transfer work

Helios Gene Gun

In these guns targeted tissue held in a vacuum, and apparatus are not very portable. The Helios Gene Gun is a hand-held device which uses low-pressure helium to accelerate DNA coated particles from the interior of a small plastic cartridge towards the target tissue. A spacer at the tip of the gun maintains optimal target distance and minimizes cell damage by venting the helium gas away from the tissue. They could not be easily used to transform target tissue that was sensitive to those conditions created by an evacuated chamber. The Helios Gene Gun can accommodate up to 12 loaded cartridges at once thus allowing for multiple firing of the device before reloading is necessary. The greatest utility of this device seems to be *in situ* transformation of animal cell and tissues (Sundaram et al., 1996).

Fig. 4.12: Microprojectile bombardment of cultured immature zygotic embryo and regeneration of transgenic plants in wheat (*Triticum aestivum*): (A) Cultured immature zygotic embryo of wheat before incroprojectile bombardment; (B) microprojectile bombardment precultured wheat embryos after proliferation and transfer to regeneration medium in absence of selective agent; (C) phosphinothricin-resistant regenerating callus obtained from precultured immature zygotic embryos bombarded with microprojectiles coated with plasmid DNA bearing a chimeric *bar* gene as selectable marker; (D) methotrexate-resistant plantlet regenerated from precultured immature zygotic embryos bombarded with microprojectiles coated with plasmid DNA bearing a chimeric *dhfr* gene as selectable marker; (E) Transformed (center) wheat platlet obtained from microprojectile bombardment using a chimeric *bar* gene as selectable marker and untransformed controls (left and right) after transfer to phosphinotricin-containing rooting (left); (F) in vitro growing methotrexate-resistant wheat plantlets from microprojectile bombardment using a chimeric *dhfr* gene as selectable marker; (G) fertile transgenic wheat plants from microprojectile bombardment to precultured immature zygotic embryos.

Factors Affecting Particle Bombardment-Mediated Transformation

1. Particle bombardment is a physical process of DNA introduction.
2. After the DNA-coated particles are introduced, there are numerous factors that influence gene expression and the subsequent recovery of transgenic plants.

3. These factors are resulting in the expression and eventual integration of the foreign DNA. For successful transformation via particle bombardment, the plasmid DNA should be carefully designed and selected, while special attention also needs to be placed on the nature, state and receptivity of the target tissues.

Components of the Introduced Plasmid DNA

The introduced plasmid DNA(s) typically contains the necessary plasmid backbone (allows growth and selection in the bacterial host), plant marker genes and the gene(s) of interest.

The *plant marker genes* can be selectable markers (which allow growth of transformed plant cells in the presence of a selective agent) and/or scorable markers (reporter genes). The genes themselves are composed of the promoter, coding region and terminator. Introns, enhancers and attachment regions are defined regions of additional DNA that can be used to enhance, modulate or stabilize gene expression.

Selectable Marker Genes

The selection of proper marker genes of choice is critical for the successful recovery of stably transformed tissues. The selectable marker allows the transformed cells to proliferate in the presence of a selective agent while nontransformed cells either do not grow or multiply at a much reduced rate. Selectable marker genes may be either antibiotic or herbicide resistance genes. The resistance genes can act either by modifying the selective agent itself to inactivate it or by producing an insensitive form of the normal target of the selective agent.

Selectable Marker Gene: NPTII

1. It is most commonly used selectable marker gene.
2. It imparts resistance to the antibiotic kanamycin.
3. The NPTII gene product is neomycin phosphorylation, which inactivates the antibiotic by phosphorylation.
4. Other antibiotics that are commonly used with selectable marker genes in plants include hygromycin, phleomycin and gentamycin.

Selectable Marker Gene : bar *gene*

1. It is most commonly used herbicide resistance marker gene.
2. It is typically used with species of the Gramineae.
3. The *bar* gene encodes phosphinothricin acetyl transferase, which inactivates phosphinothricin herbicides by acetylation.

4. The *bar* gene can be used with bialaphos of glufosinate; again, selection of the proper herbicide must be empirically determined as with antibiotics.
5. Other commonly used herbicide resistance genes impart resistance to glyphosate, sulfonyl ureas and dalapon.

Reporter Genes

1. β-glucuronidase (GUS) is the most widely used reporter gene for plant transformation.
2. The GUS gene has proven to be very useful in monitoring transient expression in a number of species.
3. Plant tissues could be evaluated at various times following bombardment by reporter genes whose expression can be monitored without causing tissue damage. Such a system could greatly reduce the effort and time involved in selection of stable transformants.
4. The firefly **luciferase gene** is a nontoxic reporter gene that has been successfully used to monitor gene expression in plants. After activity bombarded tissues are treated with luciferin (substrate) solution, luciferase activity is documented using either X-ray film.
5. Luciferase activity can also be assayed using extracts of plant tissue that are incubated with the luciferin substrate.
6. The detection techniques for luciferase severely limit the flexibility of using this reporter gene.

Introns

1. The introns are present between the promoter and coding region of a gene.
2. They significantly affect gene expression. Monocotyledonous species, in particular the Graminae, show significant enhancement of gene expression when using introns.
3. Introns may act by stabilizing mRNAs, thus leading to greater amounts of translation products.
4. Introduction of different introns between the cauliflower mosaic virus 35S promoter and GUS gene caused significant enhancement of transient gene expression in maize. The degree of expression varied depending on the intron.

Promoter Analysis

The use of reporter genes such as GUS or GFP under control of strong promoters (e.g., CaMV 35S) allows for optimization of transformation efficiency of crop plants. Particle bombardment also provides a rapid and convenient means for evaluation of promoter strength, analysis of the function of different regions of promoters, and for examination of tissue specificity of gene expression. The promoter, either in its entirety or with various deletions/mutations, is fused to a reporter gene and introduced into the target tissue. This approach has successfully been used in a number of different species.

Codon Usage

The ability to introduce genes into plants from nonplant sources had led to the realization that genes from different organisms can be processed very differently with respect to their expression when compared with have native plant genes (Diehn, et al., 1996). Unaltered DNA from nonplant sources may contain DNAs that are recognized by plants as signals for splicing or polyadenylation. Changes for codon usage resulted in a 300-fold increase in expression levels of a synthetic Bt gene.

Target Tissue

A variety of different plant tissues have been used as targets for particle bombardment-mediated transformation. Selection of the appropriate target tissue is dependent on the nature of the research. For rapid gene expression studies, various plasmid constructions can be introduced into different tissues and transient expression can be quickly analyzed to assess promoter activity without the production of stably transformed plants (IIDA et al. 1995). Almost any tissue can be used for transient expression studies as long the cell wall is penetrable by the DNA-coated particles. The choice of the appropriate target tissue, physiological state of the plant material, and pre- and postbombardment treatments are critical for the successful regeneration of stable transformants.

Embryogenic Cultures

Embryogenic cell cultures have been very useful for the production of stably transformed plants. Particle bombardment of embryogenic maize and wheat cultures was used to generate the first transgenic plants of these species. Embryogenic cultures provide a reliable source of tissue for optimization of bombardment conditions and can be used for year-round production of transgenic plants.

Shoot Apical Meristems

Shoot meristem cells are most effective target for production of stable transgenics. Theoretically, introduction of DNA into would result in the production of chimeric plants, where the transformed cells would directly give rise to germ-line tissue and the introduced DNA would be passed on to the progeny. This approach could potentially avoid or reduce tissue culture manipulations but there are many problems with shoot meristem bombardment.

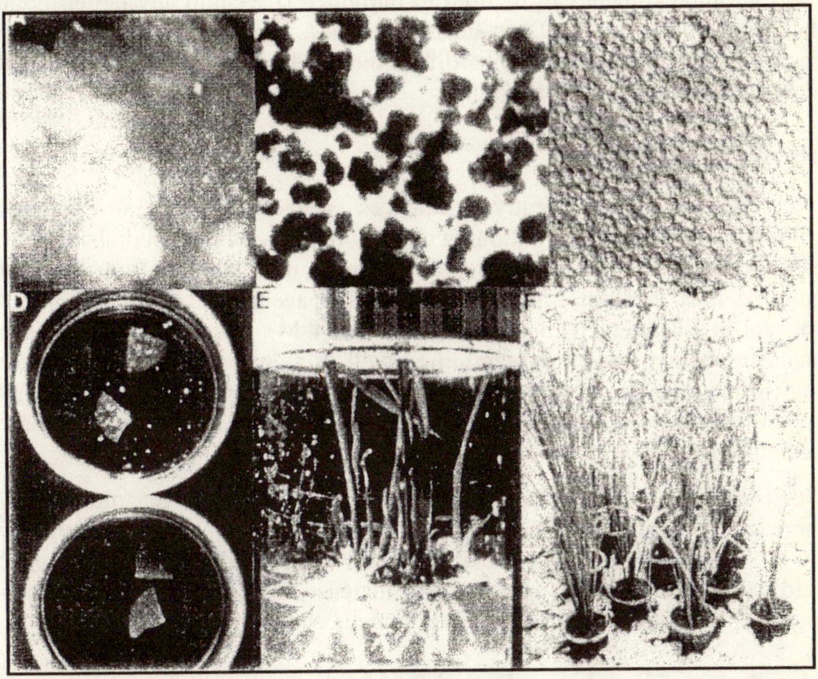

Fig. 4.13: Direct gene transfer to embryogenic cell suspension-derived protoplasts and regeneration of transgenic plants in rice (*Oryza sativa*): (A) Embryogenic ricecallus used for initiation of of suspension cultures: (B) established embryogenic cell suspension culture; (C) protoplasts isolated from embryogenic cell suspensions and treated with polyethylene glycol and plasmid DNA bearing a chimeric *hpt* gene as selectable marker; (D) selection in bead-type culture of rice protoplasts treated with (upper) and without (lower) plasmid DNA and polyethylene glycol; (E) In vitro growing plant regenerated from hygromycin-resistant callus obtained from transformed protoplasts; (F) fertile transgenic rice plants obtained from protoplasts.

In sunflower, bombarded shoot apices of particles alone (no DNA) was done and then inoculated the tissue with *Agrobacterium*. In this case, particle bombardment was used as a method to wound the target tissue for transformation via *Agrobacterium* rather than for direct DNA introduction itself. Although the resultant plants were chimeric for the introduced gene(s), it is important to note that many of the transformation events were germ-line and transgenic progeny were recovered.

Other Target Tissues

Particle bombardment can be used to introduce DNA into any target tissue that is penetrable by the DNA-coated particles. The other tissues that have been subjected to particle bombardment include leaves, pollen, styles, cereal aleurone cells and tassel primordial.

Tissue Treatment

One treatment that greatly influences gene expression is osmotic conditioning. Osmotic conditioning can be both pre- and post-bombardment. Tissues are subjected to either partial drying in a laminar flow hood or cultured in a medium containing an osmotic agent such as mannitol and sorbitol at relatively high concentration (~0.4 M). Osmotic treatment can enhances both transient expression as well as significantly increase stable transformation. The osmotica may act by causing plasmolysis of the target cells that lessens or eliminates extrusion of the protoplasm in cells that are penetrated by particles.

Other media additives are silver thiosulfate along with mannitol. They increase GUS expression in bombarded scutellar tissue of wheat, presumably by reducing the negative effects of wound ethylene. Inclusion of abscisic acid with an osmoticum (myo-inositol) increased trnsient GUS expression 1 day after bombardment of *Picea* embryogenic cultures.

The Fate of the Introduced DNA(s)

Although particle bombardment has been used for DNA delivery to plant cell for some time, it is still nuclear what happens to the majority of DNA that is introduced. Transient expression studies indicate that large numbers of cells can receive the particles resulting in approximately 10,000 GUS-positive cells per bombardment. In cells that transiently express the introduced DNA, particles are most often seen either in or directly adjacent to the nucleus.

Recombination

Plasmid DNA which is introduced into plant cells via particle bombardment undergoes recombination with other plasmid molecules as well as with the host chromosomal DNA. Recombination occurs whenever new gene arrangements are formed through exchange, elimination or insertion of DNA. Plasmid DNA can either insert cleanly into plant chromosomal DNA as a single entity or large linkages of plasmids can form and subsequently integrate (the full range of intermediate combinations is possible). Integration of a large chain of plasmids can result in a single insertion event with a high copy number of introduced plasmid DNA. With transformed event tissue obtained via particle bombardment, a single integration event rarely results in the introduction of a single plasmid. When large numbers of plasmids integrate into the same site, the plasmids may recombine with each other prior to integration into genomic DNA.

Although the exact fate of DNA introduced via particle bombardment is not clear, it appears that plasmid DNA molecules rearrange or recombine with each other to yield linkages of both whole plasmids and plasmid fragments.

Position Effect

The expression of the introduced gene(s) is dependent on the nature of the DNA as well as the final physical arrangement of the foreign DNA in the plant genome. The influence that adjacent, native plant DNA has on expression of the introduced DNA is called *position effect.*

If the foreign DNA integrates in a region of DNA that is active, the introduced gene may be expressed at relatively high levels. If the introduction occurs in a region that is not transcriptionally active, gene expression may be reduced.

Although the T-DNA from *Agrobacterium* is generally inserted into transcriptionally active regions, DNA that is introduced via particle bombardment show no known preference for insertion sites. Position effects control the variation in gene expression in different transformed clones. Methylation of introduced DNA has been shown to occur in cases where the introduced gene shows minimal activity. However, the precise association of methylation with position effects is still not clear.

Co-suppression

A reduction in expression of a gene resulting from the introduction of multiple copies of the gene is called *co-suppression.* Co-suppression is useful in some situations when a reduction in expression of a native gene is desirable. Unfortunately, co-suppression is also observed as an undesirable product of particle bombardment-mediated transformation when multiple copies of the same gene are introduced.

Scaffold Attachment Regions

To stabilize gene expression in transgene plant tissue, scaffold attachment regions (SARs; also known as matrix attachment/associated regions, MARs) have shown some utility. SARs are regions of DNA which were isolated based on their ability to bind to the nuclear scaffold. For transformation, the SARs are ligated to the flanking regions of the gene of interest. They apparently act by stabilizing or normalizing gene expression rather than enhancing overall activity (Breyne, et al. 1992). SARs from yeast (Allen, et al. 1993) and tobacco (Allen, et al. 1996) have helped to stabilize gene expression in plants while a human and soybean SAR were nonfunctional. Particle bombardment-mediated transformation efficiency could potentially be increased using SARs as SAR activity was found in transformed Booster Sequence isolated from petunia. SARs appear to be especially useful in cases of high copy number DNA introductions, which would normally result in highly variable gene expression profiles.

Targeted Recombination Systems in Plants

In order to have better control on the expression of the introduced DNA(s), it would be beneficial to predict or *direct the site integration.* This can be accomplished through either *agrolistic transformation* or homologous recombination

Agrolistic Transformation or Homologous Recombination: For agrolistic transformation, plant tissues are bombarded with plasmids that contain some of the DNA excision components from the *Agrobacterium* gene transfer system. Approximately 20% of agrolistically transformed plant cells contained precisely defined DNA inserts that would be expected for *Agrobacterium*-mediated transformation.

For targeting using homologous recombination, genes could potentially be placed in certain areas of the genome based on sequence homology between the introduced and the genomic DNA. Although targeted or homologous recombination is fairly well-characterized in some organisms, it has not been well-studied in higher plants. Very little is known about native plant recombinases with most recombination studies in plants relying on information and genes that are available from nonplant systems.

Homologous Recombination Systems: Two homologous recombination systems have been evaluated in plants.

1. *Cre-lox system* from bacteriophage P1

2. *FLP/FRT system* from yeast

In these systems, a target site containing specific DNA sequences must be introduced specific recombinase, a second piece of DNA using transformation. Following activation with a specific recombinase, a second piece of DNA replaces the target DNA through homologous recombination.

GENE TRANSFER BY MICROINJECTION TECHNIQUE

Intranuclear microinjection is a direct physical approach that overcomes many of the biological and other obstacles. Originally developed for animal cells, it is now applied to plant cells also. The microinjection technique uses a fine capillary needle to deliver DNA precisely into cells (even nuclei of cells) in such a way that the cells are not killed but survive the treatment and can develop further into clones.

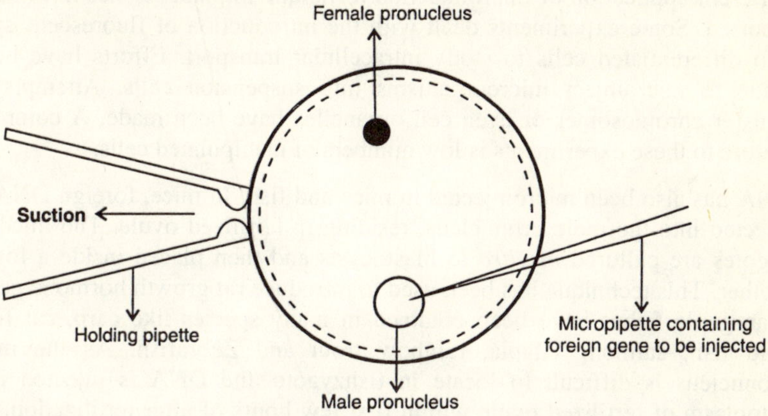

Fig. 4.14: Microinjection system.

For transformation, microinjection systems are designed for speed and operational simplicity as these factors govern the total number of cells that can manipulated. To facilitate rapid injection rates, protoplasts with partially reformed cell walls can be attached to solid support (glass coverslips or slides etc.) with poly-L-lysine or artificially bound substrate without damaging the cells. After one day in culture, the cell walls of protoplasts are firm enough to provide support and protection yet they can be easily penetrated with micropipettes worth 0.5-10 μm diameter tip). Each protoplast is microinjected, segregated and transferred to optimal condition to obtain transformed cell line. An alternative method employs a blunt holding pipette to hold each protoplast under suction and segregating it directly into microdrop culture after microinjection.

Several advantages are associated with the microinjection technique. First of all, the amount of DNA delivered per cell is not limited by the technique and can be optimized. This improves the chance for integrative transformation. Secondly, the delivery is precise, even into the nuclei of the target cell, again increasing the chance for integrative transformation. Thirdly, the small structures can be injected containing only a few cells and with high regeneration potential, for example, microspore derived and zygotic proembryos. Lastly, it is a direct physical approach and therefore host-range independent.

However, there are some major drawbacks like, only one cell is targeted per injection and the handling requires specialized skill and instrumentation.

Applications of Microinjection Technique

Different application of microinjection technique in plant science have been reported. Some experiments dealt with the introduction of fluorescent dyes into differentiated cells to study intracellular transport. Efforts have been made to microinject microorganisms into suspension cells. Attempts to transfer chromosomes or even cell organelles have been made. A common feature to these experiments is low numbers of manipulated cells.

DNA has also been microinjected in mice and fish. In mice, foreign DNA is injected into the mole pronucleus, residing in fertilized ovule. The injected zygotes are cultured *in vitro* to blastocysts and then placed inside a foster mother. This technique has been used to introduce rat growth hormone gene. Transgenic fishes have been obtained in many species like carp, cat fish, gold fish, salmon, Tilapia, rainbow trout and Zebrafish. As the male pronucleus is difficult to locate in fishzygote, the DNA is injected into cytoplasm of fertilized ovule within first few hours of after fertilization. As the fishes undergo external fertilization, transfer of embryos into foster mothers is not needed. Transformed embryos are instead hatched in trays. Various gene transferred into fish are human or rat growth hormone, chicken delta-crystalive protein, *E. coli* β-galactosidase gene, *E. coli* neomycin resistance gene, *E. coli* hygromycin resistance gene, winter flounder anti freeze protein (AFT) gene, and, cDNA of rainbow trout growth hormone. Cytoplasmic microinjection is a tedious and time consuming procedure and is unsuitable for mass gene transfer.

The Technique

The macroinjection technique involves the use of injection needles with diameters far wider than cell diameters for delivering DNA into wound sites within tissues. However, at the same time it destroys those cells which are in direct contact with the injected DNA. To integrate, the DNA must gain entry to intact cells adjacent to the wound site.

There are a variety of problems associated with experiments that require DNA passage through single, double (in the case of pollen) or even many cell walls. The uptake of DNA across cell walls is a very rare event. Majority of DNA molecules applied get attached to cell walls before they have a chance to enter the cell. Penetration is further reduced by subsequent presence of cell wall. Moreover, DNA does not normally move from cell to cell. Thus, recombination occurs at relatively low frequency by this method. Microinjection has been mostly applied to ovule, embryos and other tissues

Lipofection

The use of liposomes for transformation of transfection system is called lipofection. Liposomes, as defined by Bongham, in 1965, are special type of lamellar phase in which water is self contained and the lipid molecules are disposed in bimolecular layers attached by their non-polar interfaces.

Liposomes possess many properties akin to those of biological membranes. They are easy to manipulate, their lipid contents can be varied at will and many substances may be trapped inside the interlamellar spaces. They are found to be ideal carrier systems and are now used in genetic engineering to deliver foreign DNA molecules into a desired cell.

DNA or RNA, encapsulated in liposomes, can be transferred into plant protoplasts by either direct fusion with the naked plasma membrane or through endocytosis of the liposome. Although it is efficient in terms of DNA encapsulation and transfer, the method once again requires protoplasts and its application is limited to species for which protocols are available for protoplast regeneration. Foreign macromolecules, for example impermeant fluorescent dyes, RNA and DNA, can be introduced into walled plant cells by an alternative strategy which involves encapsulation of the desired probe into liposome that can be pressure-injected into the vacuole. There are two advantages of this method. First of all, the liposome protects the DNA from being damaged by the acidic pH and proteases present in the vacuole. Secondly, when liposomes are constructed from material that is compatible with the lipids of the tonoplast, liposome-tonoplast fusion occurs (Madore *et al.*, 1986) and so the encapsulated DNA should be successfully delivered into the cytoplasm.

Liposomes would appear to offer the following advantages as vectors for gene delivery into plant cells:

1. Enhanced delivery of encapsulated DNA by membrane fusion.
2. Protection of nucleic acids from nuclease activity.
3. Delivery into a variety of cell types besides protoplasts entry through plasmodesmata.
4. Delivery of intact small organelles.
5. Protect the genetic material from immunogenic reaction (liposomes are non-immunogenic).
6. Composition of lipid can be manipulated to confer specific properties to liposomes.

The use of liposome for delivery of gene was first described for delivering polio virus RNA. Later liposomes were used in combination with polyethylene glycol for delivery of plasmid DNA into tobacco protoplast.

Silicon Carbide Fiber-Vortex

A procedure has been recently developed for the injection of DNA into *Zea mays* and *Nicotiana tabacum* tissue cultures using silicon carbide fibres. This method involves a mixture of plasmid DNA encoding a selectable or screenable marker gene, silicon carbide fibers, and tissue of the species to be transformed in an Eppendorf tube. Mixture is vortexed to facilitate cellular penetration by enhancing frequent and forceful intercellular collisions. DNA deliver in this system is presumably due to cell wall penetration by DNA coated silicon carbide fibers. The silicon carbide fibers apparently act as microinjection needles facilitating DNA delivery.

Advantages of the silicon carbide fiber-mediated transformation method over other procedures includes the ability to transform walled cells (thus avoiding protoplast isolation), relative ease of the procedure, and very low equipment costs.

Sonication

The latest method developed uses ultrasound to facilitate the uptake of nucleic acids into plant cells. By this technique, plant protoplasts and cells in suspension and intact pieces of plant tissue can readily be transformed. The protoplasts, suspension cells or small pieces of tissue are suspended in a few ml of sonication medium in a microcentrifuge tube. Plasmid DNA (and possibly carrier DNA) is then added and after rapid mixing the samples are ready for sonication.

The pulses of ultrasound are delivered by ordinary machines. The microtip (normally tapered) is immersed 2-3 mm into the suspension and pulses of selected intensity and duration are delivered. The cells are finally transferred to fresh growth medium.

There are two possibilities for permeabilization of cell and uptake of DNA. Firstly the violent collapse of activation bubbles, generating high-pressure and high-temperature shock waves could possibly cause localized rupture of the plasmalemma and lead to uptake of exogenous solute. This may be followed by re-establishment of membrane integrity. The second hypothetical possibility may be the existance of a critical hydrostatic pressure at which the intrinsic membrane potential is sufficiently high to induce mechanical breakdown of the membrane (60-100 MPa) in a hyperbaric chamber. Consequently, it is possible that either the high oscillating pressure generated by the ultrasonic field and/or the high-pressure shock waves orginating from collapse cavitation could produce such higher hydrostatic pressures that reversible membrane breakdown would occur.

This method has been applied for transferring plasmid DNA into protoplasts, suspension cells and tissues of sugarbeet and potato.

ELECTROPORATION

This method uses an electrical pulse to introduce DNA into cells. By exposure to brief electric shock, the cells become temporarily permeable to DNA. If the required DNA fragment is kept in direct contact with the protoplast membrane, the entry of DNA molecules into the cells is facilitated. Successful DNA transfer through this technique has been shown in bacterial, yeast, fungal, plant and mammalian cells. Among plants, successful gene transfer using electroporation method has been reported in protoplasts of tobacco, petunia, maize, rice, wheat and sorghum. In carrot, maize and tobacco protoplasts, bacterial CATgene with different plant promoters has been introduced and efficiently expressed. Electroporation is applicable to both monocot and dicot protoplasts.

It has been recently hypothesized that the mechanism of DNA entry into the cell during electroporation may be either due to unspecific membrane process, or it may be specifically mediated by permease. The activation of permease may be due to high electric field, and thus, may lead to thinning of membrane. Hole may be created as a result and stabilized by a favourable dipole interaction with the electric field. Various advantages of electroporation have been listed. It is a convenient, simple, fast and efficient method applicable to a wide range of plant protoplasts and involves low cell toxicity.

The main disadvantage of electroporation is its requirement for protoplasts and therefore a system where regeneration of these protoplasts to calli or whole plants have been established. Consequently, it may not be applicable to some plant systems. Recently an electroporation system using organized and easily regenerable tissues such as immature maize embroys has been developed.

POLY-CATION MEDIATED DNA UPTAKE

Direct DNA uptake by protoplasts can be stimulated by using hydrophilic, long chain polycations such as polyethylene glycol (PEG), poly L-lysine, poly L-ornithene or DEAE-dextran with or without metal ions such as Zn^{2+}, Li^{+}, Cs^{+}, Rb^{+}, K^{+}, Na^{+}, Ca^{2+} or Mg^{2+}. The technique is efficient and can be applied virtually to every protoplast system and for transfer of even small quantities of DNA.

Polyethylene-glycol is an extremely hydrophilic molecule and it removes 'free water' when placed in a solution. Free water indicates such molecules, which interact with charged (usually ionic) molecules soluble in water.

Thus, under high PEG concentration (15-25%), ionic macromolecules such as DNA can no longer stay in solution and get precipitated. The protoplast membrane as well as DNA is normally negatively charged, and therefore, charge repulsion is obvious. This repulsion limits the interaction between DNA and protoplasts. At very high concentration, PEG seems to minimize charged repulsion effects thus stimulating DNA uptake by endocytosis without any gross damage to protoplasts. Studies have been conducted with *Petunia, Nicotiana*, rice, maize etc. In these experiments, PEG was used in combination with pure Ti plasmid or calcium phosphate precipitated Ti plasmid mix with a carrier DNA. Mechanism of cross membrane transport in presence of PEG is still to confirm.

DNA-uptake may be enhanced as a result of the interaction of the polycations at one or more of the following levels:

1. Protection of the exogenous DNA against nuclease degradation.

2. Increase in the permeability of the plasma membrane.

DNA transformation by using the polycation "polybrene" (hexamidethrine bromide) is new, although similar techniques have been used for several years in the transformation of animal cells. The major advantages of the polybrene technique are the following:

1. It is less toxic than other polycations

2. The high transformation efficiency allows very small quantities of plasmid DNA to be used, thus it is relatively easy to introduce only one sequence of a genome. Even with quite high plasmid concentrations only one or a few copies of the plasmid are integrated and concatamers of integrated plasmid are rarely observed.

3. Since no carrier DNA is used to integrate plasmid DNA into the host genome thus, the sequences surrounding the site of integration can be analysed.
   (a) The transformation frequency appears to be independent of nature of plasmid (supercoiled or linear).
   (b) The polybrene method can be used for both stable and transient expression.

CA-DNA CO-PRECIPITATION METHOD

The basis of the Ca-DNA technique is the formation of a co-precipitate of plasmid DNA with calcium phosphate, formed by adding $CaCl_2$ to a solution of DNA in phosphate buffer. On contact with protoplasts, the co-precipitate is transferred across the plasma membrane in a calcium requiring process; and in order to encourage the endocytotic uptake of the precipitate, the protoplast/Ca-DNA complex is treated with polyvinyl alcohol and high pH calcium. This method was first used for plant transformation by Hain *et al.*, (1985) using a tobacco leaf protoplast system.

SILICON WHISKERS

Silicon carbide forms microscopically small needles (silicon whiskers) that can be used to perforate plant cell walls and plasmalemma of suspension cells or protoplasts. Technically, the methods is very simple: a mix of whiskers and plant cells in a small test tube is shaken. Thus, only Eppendorf tubes and a vortex is sufficient equipment. This opens the way for DNA diffuse into the cells (Kaeppler et al., 1992). Although, details of the mechanisms by which silicon carbide wiskers work is unclear, stable transformation with maize has been recently published (Frame et al., 1995). The overall efficiency of the wiskers approach is between one-fifth and one-tenth that of conventional microprojectile shooting, but is advantage would be its "low tech" character, dispensing with any sophisticated or expensive instrumentation. However, currently, only maize has been transformed and a highly regenerating system was used. It remains an open question whether the whiskers approach will be useful in organized tissue.

ULTRASONICATION

Gene transfer by ultrasonication is the latest and simple method, where, plant protoplasts, cells in suspension and intact pieces of plant tissues can be transformed. The target cells are suspended in sonication medium in a microcentrifuge tube. Carrier DNA is added and microtip (generally tapered) is immersed in the sonication medium. Pulse of selected intensity and duration are delivered by the ordinary homogenizing machines. Transformed cells are finally transferred for selection.

There are two hypotheses for permeabilization of membrane by sonication method. First assumes that reversible localized rupture of plasmalemma results in is due to violent collapse of activation bubbles which results in generating high-pressure and temperature shock waves to plasmalemma. The second hypothesis predicts the existance of critical hydrostatic pressure responsible for reversible mechanical breakdown of membrane.

This method has been used in tobacco, wheat and sugarbeet. In tobacco, transgenic plants were obtained at 22% frequency with ultrasonication. Ultrasound transformation is unique among existing methods because of its versatility as plant protoplasts, cell in suspension and intact pieces of plant tissue can be readily transformed. This feature, alongwith the fact that the equipment for ultrasound transformation is simple, cheap and multifunctional should allow for substantial future applications.

U.V. LASER MICROBEAM

A new method of introducing exogenous gene materials into cells is laser microbeam irradiation. DNA can be incorporated into whole plant cells and has been successfully applied to tobacco. DNA enters the cell through a self-healing hole made with a laser microbeam.

There are several advantages to this method. A laser beam can be used inside of tissues, cells and organelles, thus increasing the rate of transformation and the speed of operation. Wide controllability of the laser beam intensity, flexibility in optical manipulation wide choice for laser wave length and pulse width are other advantages. It is also applicable to cells with hard walls. Its wide application in the field of cell biology is the ability to transfer cytoplasmic material. Finally, future versions of this system are expected to be more automated which should greatly increase output as well as the success rate.

This technique can be used for genetic manipulation of plant tissue, individual cells, protoplasts and organelles. Furthermore, it permits selective fusion in a of part of a subcellular organelle present inside a single living cell.

A number of techniques exist today for transferring genes into plant protoplasts and cells. Most of these techniques use protoplasts because the cell wall is not sufficiently permeabilized for DNA uptake. As protoplasts of many species are often difficult to regenerate efficiently into fertile plants it can not be applied to any system. These techniques can be preferably used as an analytical tool, for example, in testing the function of gene constructs.

Moreover, most of the systems require sophisticated and costly equipment. In addition, it does not allow the combination of both accurate quantitative estimates of gene expression and efficient generation of transgenic plants.

However, successful gene transfer has been reported in few systems using above techniques, and much research in the context is still in progress.

EXPRESSION OF INDUCED GENES

Genetic Engineering in plants includes formation of recombinant DNA, its introduction into cells or protoplasts and more importantly expression of the foreign genes. After introduction of foreign gene, complete plants are regenerated from transformed cells or tissues. Regenerated plants are then examined for expression of gene constructs. It is a lengthy process, involving selection and bulking of tissues for analysis of integrated genes. This duration can be avoided by studying the 'transient expression' of the constructs introduced directly into protoplasts. In this approach, foreign DNA is added to protoplasts by gene transfer method and these protoplasts are cultured for 1-2 days. The protoplasts are then harvested and assayed for products encoded by the introduced (but not necessarily integrated) genes. This confirms the incorporation of required gene into the protoplast.

Reporter Gene for Transient Expression

A reporter gene is added to gene of interest before its entry into the target cell. With the help of reporter genes, information on aspects of gene constructs such as the relative-level and tissue specificity of expression can be obtained.

Reporter genes are frequently introduced into genetically modified plants for identifying and tracking genetic modification as they tag the transformed cells. They find more use where the genetically modified plants can not be regenerated from single cell and direct selection is not feasible or effective.

The important features needed in reporter gene include its low background activity in plants and absence of any detrimental effect on metabolism. The gene product should have moderate stability *in vivo*, so that down regulation of gene expression as well as gene activation can be detected.

β-Glucuronidase Gene

The β-Glucuronidase (GUS) gene, encoded by the Uid A locus of *E. coli*, is most popular reporter gene used in plant transformation. The product of this gene i.e. enzyme β-glucuronidase breaks down glucuronides giving a coloured reaction. A blue coloration develops in the cells due to the reaction, thus expressing the gene. GUS activity can be quantified by excitation with light at 360 μm and monitoring fluorescence at 460 μm. Its presence can be detected *in situ*, inside the plant tissue, used either as thin section or in any other form. Several glucuronides which can be used as substrates, include β-nitrophenyl glucuronide (PNPG), 5-bromo, 4-chloro, 3-indolyl glucuronide (BCIG), naphthol As-BI glucuronide (NAG) and resorufin glucuronide (REG).

The major advantage of using this reporter gene lies in its assay, which is easy and sensitive. The expression of GUS gene fusions can be easily quantified by fluorometric assay, and histochemical analysis can be used to localize gene activity in transgenic tissues.

Luciferase Gene

This gene can be used as reporter gene and for transient gene expression. The activity of luciferase gene fusion can be assayed in transformed plants by their ability to grow in dark. The enzyme luciferase differs in bacteria and *Photinus pyralis* (firefly). While in bacteria, it consists of two peptide subunits coded by gene lux A and lux B, in case of firefly it consists of single polypeptide coded by a gene lux. The enzyme luciferase uses different substrates for luminiscence in these two systems. When the source of the gene is bacteria, the substrate used is an aldehyde (decanal) which is supplied exogenously with FMNH2. Alternatively when the source is firefly, luciferin (a heterocyclic carboxylic acid) is used as a substrate in the presence of ATP and oxygen. It generates oxyluciferin and yellow green fluorescence.

Chloramphenicol Acetyl Transferase Gene

This gene, isolated from *E. coli*, is primarily used as reporter gene and not as marker gene. As this gene is not present in mammals and higher plants, gene construct introduced in them can be detected by chloramphenicol acetyl transferase (CAT) assay. The enzyme uses acetyl Co A and, chloramphenicol (32P) as substrates and helps in the transfer of acetyl group to chloramphenicol. Presence of acetyl-chloramphenicol can be detected through autoradiography. CAT gene is also used for identification of a number of regulatory sequences.

Nopaline (nos) Synthase Gene

The disarmed Ti-plasmid obtained from *Agrobacterium tumefaciens* is used as a vector for gene transfer in dicot plants. The disarmed plasmid is without tumour causing region of TDNA. Hence it can not cause tumour. However, the border sequence and the nos gene are retained e.g. pGV3840. This plasmid is derived from the Ti-plasmid pTiC58. The nos gene codes for the enzyme nopaline synthase. Plant cells transformed using such a plasmid (pGV3840) will produce nopaline in their cells. The presence of nopaline can be identified by biochemical analysis. A positive result will show that cells are transformed.

Neomycin Phosphotransferase Gene (NPT II)

This gene is used both as selectable markar gene as well as reporter gene. This is the first marker to be used and is isolated from *E. coli.* This gene has been successfully used in many systems including mammalian cells and fungi. In solanacous plants such as petunia, tobacco and tomato, it provides an excellent, unambiguous selectable marker. For assay of the gene product (i.e. neomycin phosphotransferase), this enzyme is first fractionated using non-denaturing polyacrylamide gel electrophoresis (PAGE). Since the enzyme detoxifies kanamycin by phosphorylation, radioactively labelled ATP (^{32}P) is also used with it and incubated at 35°C. The phosphorylation leading to incorporation of ^{32}P in kanamycin can be detected by autoradiography. Another method of enzyme assay is much easier. Dot blots of cell extracts are obtained on filter and this filter is incubated with substrate and finally autoradiographed to detect NPT II activity.

Anthocyanins

It utilises the C1, B and R genes, which code for trans-acting factors that regulate the anthocyanin biosynthetic pathway in maize seeds. The introduction of these regulatory genes into plants by microprojectile bombardment induces cell-autonomous pigmentations in non-seed tissues.

Dihydrofolate Reductase Gene

Another selectable marker is the mouse gene for dihydrofolate reductase. It encodes an altered protein that does not bind to methotrexate, a potent inhibitor of this essential enzyme. Petunia cells that contain and express this gene are able to grow on concentrations of methotrexate that completely inhibit the control cells.

For expression of these reporter gene, suitable promoters are added adjacent to the former.

Promoter Elements

One of the best studied plant promoters is the 35S promoters. This viral promoter produces the cauliflower mosaic virus 35S genomic RNA. In the early experiment approximately 1000 bp of promoter DNA were fused to various reporter genes. Analyses of protoplasts, transformed calli and transgenic plants demonstrated that the reporter genes with promoters were always expressed at high levels and were also insensitive to various endogenous and environmental cues (such as hormones, heat shock or light). The so called constitutive expression made the 35S promoters popular in dicot species. In monocot cells, the expression levels with the 35S promoter are 10 to 100 times lower than in dicot. A variety of promoters and other methods have been used to increase the level of expression of genes introduced into monocot cells. The incorporation of an intron into the transcriptional unit of the foreign gene increases gene expression in transformed monocot cells. The combination of the first intron of maize alcohol dehydrogenase 1 gene (Adh 1) and the CaMV 35S promoter have been used to control expression of the bar gene in transformed maize, oat and wheat. An alternative strategy is to include a modified maize Adhi sequence in the Emu promoter. This has been used in expression of neomycin phosphotransferase (npt II) gene in transformed rice and sugarcane. Finally, efforts have been made to isolate monocot promoters that show high level constitutive activity. The promoters of the genes for rice actin (Acti) and maize ubiquitin (Ubii) have been used to express the bar gene in transgenic wheat, barely and rice.

The direct transformation systems are most suitable for transient expression studies at present. However, the most successful and widely used system for stable transformation of plant is that based upon the Ti plasmid of Agrobacterium. In this method, the tissues are inoculated with Agrobacterium carrying the requisite vector having gene of interest. The next step is to select the transformed cells or tissues. This is facilitated by the presence of selectable marker genes available in vector.

Marker Gene

A good selectable marker gene is one that provides resistance to a drug, antibiotic, or other agent that arrests the growth of normal plant cells (plant cells not harboring the marker gene). The selective agent should inhibit the growth of plant cells or kill them slowly. Agents that kill cells rapidly often result in the release of phenolic compounds and other substances (from the dead cells) that are toxic to the growth of the remaining, otherwise resistant cells. Various markers have been used for selection of transformed cells

Three marker genes among them are used more extensively, providing resistance to the antibiotics:

1. Kanamycin (and the related aminoglycoside G418,which are more effective growth inhibitor in eukaryotes),
2. Hygromycin, and
3. Methotrexate

Transformed system (tissues/organ) is placed on regeneration medium containing an antibiotic like **kanamycin** and transformed plants can be obtained directly. Cells, which are not transformed, die due to kanamycin. The heterogenous explant like calli or non-transformed tissue may escape initial selection. The leaf disc instead, exihibits the regeneration from edges, which is in contact with selective agents and thus, is a better alternative.

Inactivation of Foreign Gene Expression

There are a number of conditions that can interfere with gene expression in transgenic plants. These include:

1. Poor recognition of promoter elements and differences in transcription efficiency between the source of the promoter and the transformed material.
2. Inefficient mRNA termination, polyadenylation, processing and/or stability.
3. Inefficient translation due to low rates of translation initiation and/or inappropriate codon usage.
4. Poor design of expression vectors, which might contain multiple repeated regions or multiple genes with opposite orientations leading to transcriptional interference.
5. Antisense-mediated gene inactivation.
6. Gene silencing by homologous sequence.

Transgene Inactivation and Methylation

Multiple copies of the transgene also suppress their expression. There is reverse correlation between gene copy number and expression level. Low expression is correlated with increased methylation of the transfer DNA. It is well established that methylated DNA is not or poorly transcribed. The most common DNA modification in plant cells is cytosine methylation at CG dinucleotides and CNG (were N can be any base) trinucleotides. Methyl-cytosine can interfere with protein-DNA interactions. This phenomenon has been observed with maize dihydroflavonol-4-reductase (DFR) gene in Petunia and with GUS transgene in tobacco.

Double Transformations and Gene Silencing

Introduction of a second T-DNA construct into tobacco already containing T-DNA also lead to the suppression of genes located on T-DNA first. Presence of second T-DNA leads to methylation of the genes of T-DNA first, thus resulting in their suppression. If T-DNA second is crossed out, the T-DNA first genes become demethylated and reactivated. It has been suggested that homology between the DNA sequences of the two T-DNAs, even though they are not transcribed, triggers the silencing.

Post-Transcriptional Gene Suppression

Supression of β-1-3-glucanase and chitinase transgene expression is only observed in plants that are homozygous for the transgene. The β-1-3-

glucanase transgene suppression is the result of a post-transcriptional process and is observed only in homozygous and in haploid plants.

A number of issues still need to be addressed before transformation technology can be exploited fully in the agronomic improvement of plants. These include those that act to regulate the efficiency of transcript termination, transcript stability, post-transcriptional modification and translation efficiency. An increased understanding of these regulatory components in plants will prove profitable for achieving the regulated expression of foreign genes in transgenic plants.

GENETIC TRANSFORMATION IN MONOCOTYLEDONOUS PLANTS

The initial development of system for genetic transformation of monocotyledonous was slow. Formation of tumors and production of opines in certain monocotyledons suggested that at least some monocotyledonous species might be susceptible to infection by *Agrobacterium tumefaciens* (Ti-plasmid).

Table 4.2: Genetic Transformation in monocotyledonous plants

Work	Reference
Transformation of monocots and cereals	Heii et al. (1997)
few transgenic rice plants by inoculating immature embryos with a strain of *A. tumefaciens*	Chua et al (1993)
Reported a method for efficient production of transgenic rice plants from calli of japonica cultivars that had been co-cultivated with *A. tumefaciens*	Hiei et al (1994)
The successful application of calli of japonica cultivars method to Basmati cultivars of indica rice	Rashid et al (1996)
Achieved efficient transformation of maize by *A. tumefaciens*	Ishida et al. (1996)

Factors involved in *Agrobacterium*-mediated transfer of genes to monocots

1. This work is mainly done on cereals (particularly rice).
2. Induction of *vir* genes, one of the most commonly used techniques in the transformation is the addition of phenolic compounds. such as acetosyringene to co-culture or bacterial cultures.
3. The cell division should be active.
4. Bacterial strains and vectors
5. Composition and quality of the culture medium.

6. High osmotic pressure.
7. Host genotypes, and
8. Acidic pH.
9. Selectable marker
10. Temperature of the room below 28° C.

Direct Gene Transfer Monocotyledonous Plants

These method are used for direct DNA transfer in monocots(cereals). Various methods are :

Polyethylene glycol (PEG) medicated gene transfer

PEG acts as an important fusion agent in somatic cell hybridization. It is being used in obtaining transformed cells by delivering DNA into protoplasts (Davey et al., 1980; Krens et al., 1982). The foreign DNA in the form of plasmid DNA is inserted in plant protoplast resulting in the recovery of fertile regenerated tobacco (Paszkowski et al., 1984).

Electroporation

Electroporation means the use of high voltage electrical pulses to a solution containing a mixture of protoplast and foreign DNA. The pulses reverse pores created in membrane by the short electrical pulses. Transgenic plants of maize have been obtained by this method(Rhodes et al., 1988).

Microinjection

It is the injection of DNA into plants cells including cultured cells and intact plant organs to produce transgenic plants (Neuhaus et al., 1987; Dela Pena et al., 1987). For example, selected embryoids derived from microscope culture of *Brassica* are transferred for microinjection to a minute droplet of culture medium (Neuhaus et al., 1987). Embryos were injected with **NPT- II** gene constructs. Injected DNA is considered to be transported by plants vascular system to the germ line.

Micro projectile bombardment

This method helps in delivery of DNA into cells by projective bombardment. Micro projectiles (small high density particles) are accelerated to high velocity by a particle gun apparatus. These particles acquire sufficient kinetic energy to penetrate cell membrane and carry foreign DNA into bombarded cells. Klein et al. (1987) used this method for *Allium cepa* epidernal tissues. Treated tissues showed varied levels of chloramphenicol acetyl transferase (CAT).

PROBLEMS INVOLVED IN GENE TRANSFER

Expression System

Adequate expression of gene is essential with high expression at right time and in a particular tissue. For high expression strong promoters like 35-S promoter of Cauliflower Mosaic Virus (CaMV) is used which has a wide host range. More specific promoters are required for tissue specific attachment regions (SAR's) are used to increase expression. SAR's act by forming loops in chromatin structure thereby reducing gene silencing in cells having limited number of transgenes. But high amount of m-RNA may not have corresponding amount of product.

Cellular Localization

Abiotic stress causes injury to membrane (by formation of reactive oxygen intermediates) affecting membrane bound process finally influencing metabolism of plant. It is desirable to locate membrane modifying, osmoprotectant producing or radical scavenging enzymes to defined organelles. Transit peptides are required to ensure correct localization of enzymes expressed. Typical transit sequence used to direct polypeptides to chloroplast is from small subunit of Rubisco (Schreier et al., 1985; Allen et al., 1996).

Proteolysis, Protein Folding and Prosthetic Group Acquistion

Short polypeptides after expression may undergo proteolytic degradation. This can be avoided by processing and folding of proteins by co-expression of chaperon proteins in transgenic plants especially in case of large enzymes. Oxygen radical metals like, Fe, Zn, Cu, Mn, Therefore, the metal ions should be present in adequate amount.

Precursor Availability

Amount of precursor has a great effect on product formation. Precursor can be extracted from environment or synthesized in cell from existing metabolic pathways.

Inhibitory Environments

Intracellular condition in transgenic plants (pH, temperature, ionic strength) affect foreign proteins. It is usually operated during varied environmental conditions. Therefore, the introduced gene should be taken from closely related plant.

Side Reaction of New Compounds

Endogenous enzyme activities may deplete the product pool or form toxic compounds. Therefore, analysis of transgenic plants using NMR is important for detection of any metabolic changes.

5

Expression of Induced Genes

Genetic Engineering in plants includes formation of recombinant DNA, its introduction into cells or protoplasts and more importantly expression of the foreign genes. After introduction of foreign gene, complete plants are regenerated from transformed cells or tissues. Regenerated plants are then examined for expression of gene constructs. It is a lengthy process, involving selection and bulking of tissues for analysis of integrated genes. This duration can be avoided by studying the 'transient expression' of the constructs introduced directly into protoplasts. In this approach, foreign DNA is added to protoplasts by gene transfer method and these protoplasts are cultured for 1-2 days. The protoplasts are then harvested and assayed for products encoded by the introduced (but not necessarily integrated) genes. This confirms the incorporation of required gene into the protoplast.

PLANT GENE EXPRESSION SIGNALS AND GENETIC MARKERS

Expression Signals

Foreign genes are introduced into plants to alter the properties of the recipient. This is achieved by either expressing a completely new gene or by altering expression characteristics of an endogenous gene. In both instances, a protein or RNA product has to be produced from the transformed DNA in sufficiently high amounts and at the proper time and place. A prerequisite for proper gene expression is the presence of *cis*-active signals within the transgene that's in association with cellular factors, cause production of the RNA (*transcription*), RNA maturation (*splicing*, 3' –end formation), transport of the RNA to the cytoplasm, *translation* of the RNA into a protein, and modification and transport of the respective proteins. Here, we will give a brief review of signals involved in RNA and protein production that have been used to express foreign genes in plant cells and plants.

Constitutive Promoters

Transcription initiation RNA polymerase II-dependent genes is largely contrlled by sequences upstream from the transcribed region. These promoter sequences are the main determinants of quality and temporal and spatial control of RNA production. A large number of plant genes have been isolated, and for many of these, promoters have been partially characterized. In general, all promoters consist of a mosaic of different sequence elements, the interaction of which with specific cellular factors results in the particular expression characteristics. For most of the early applications of plant genetic engineering, promoters have been used that cause expression in all or almost all tissues of a transgenic plant and that are independent of development or environmental stimuli. The most widely used of these constitutive promoters is the 35S promoter of the paraetrovirus cauliflower mosaic virus (CaMV). Other constitutive dicot promoters are the CaMV 19S promoter and the *nos, ocs,* and *mas* promoters derived from *Agrobacterium tumefaciens.* These promoters are about 50-fold weaker than the 35S promoter, and the activity of the agrobacterial promoters is influenced by hormones and wounding. Agrobacterial promoters are mainly used in dicots, although activity for the *nos* and *mas* promoter in monocots has been described.

Useful future alternatives to the CaMV 35S promoter could be corresponding promoter of figwort mosaic virus or a strong constitutive dicot promter from an *Arabidopsis thaliana* ubiquitin gene that have been described as strong, constitutive dicot promoters, but are as yet not used frequently for purposes of genetic engineering. For monocots, the CaMV 35S promoter is usable, although its activity is relatively low.

Tissue-Specific and Inducible Promoters

When transgene expression should be directed to certain tissues or time points during plant development, regulated promoters are required. A great number of tissue-specific genes have been characterized from many different plant species. Again specificity and amplitude of gene expression is controlled by an interplay of *cis-* and *trans-*acting elements. Promoter elements conferring specificity for seeds, tubers, vegetative organs, and green tissues have been studied most extensively. Sequence elements controlling specificity are apparently quite conserved; therefore, such promoters can often be used in heterologous systems either as complete promoters or as isolated elements fused to an unspecific core promoter. A variety of such promoters have been used to specifically express transgenes in homologous or heterologous plants. Particular targets have been the storage tissues in seeds (e.g., using the napin gene promoter of *Brassica napus*) or the 2S albumin gene promoter of *Arabidopsis thaliana* and tubers (promoter of a potato patatin gene and the vegetative organs (tapetum-specific promoter from tobacco. For gene expression in monocot seeds, the promoters of a maize zein gene and a rice glutelin gene may prove useful.

Inducible promoters, in principle, offer the possibility to express a transgene only in response to an external stimulus. However, natural-inducible promoters are often also under developmental control, which may result in unwanted expression of a transgene during plant growth. Likewise, usually a whole group of promoters will respond to a given natural stimulus. Artificial, inducible promoters have been developed using bacterial repressor operator systems. But, so far, promoter induction requires the removal of a repressing compound and, therefore, is not useful for applied purposes outside the laboratory.

Transcription Terminators and Polyadenylation Signals

Efficient RNA production involves not only transcription initiation, but also its termination. Generation of a functional RNA 3' –end requires the presence of a processing signal (called terminator or polyadenylation signal). Very few plant polyadenylation signals have been characterized to any extent. No species nor tissue specificity has yet been reported in plants.

Model signals contain an AATAAA sequence 10-30 nts upstream from the polyadenylation site and a GT-rich region upstream or downstream. The AATAAA sequence is not present in all signals and, even if present, may not be important for polyadenylation. The most frequently used signals are derived from CaMV or the *nos* or *ocs* genes of the agrobacterial T-DNA and are used as about 200- to 500-bp-long DNA fragments.

Introns

Primary RNA transcripts usually contain introns that have to be removed by splicing. In maize cells, gene expression from the CaMV 35S promoter or the maize *adh1* promoter could be enhanced up to 100-fold when introns from maize *adh1*, *bronzel,* or *shrunken1* genes were inserted. These introns were also effective in wheat and rice. Other monocot introns with stimulatory effect are derived from the rice *actin1* gene, and a maize ubiquitin gene. Although stimulatory effects were also obtained with the first intron of a castor bean catalase gene in rice, monocot introns in dicot plants are usually inhibitory because splicing is inefficient.

Translation

The main determinant of the competitiveness of an mRNA for ribosomes seems to be the structure of the RNA 5' –end. Unstructured ends are thought to allow a more efficient binding of 43S ribosomal subunits to the capped RNA end and their subsequent migration to the start codon. Because the start codon is almost always approached by linear scanning, the absence of possible alternative translation initiation sites upstream from the protein-coding sequence is also important for efficient expression. Translation initiation efficiency is influenced by the sequence context surrounding the ATG start codon. A survey of plant gene sequences suggested on optimal context of AACA*AT*GGC. The most important bases are located at positions –3 and +4 have been verified by mutation analysis in transgenic plants and in protoplasts. However, the importance of an optimal consensus seems to be variable (two- to eightfold) and, under natural conditions, differences is start codon usage may occur that are not explainable by context rules. It has been speculated that in plants as in other eukaryotes, translation may also be influened at the level of elongation (e.g., by the presence of clusters of codons for which only low levels of charged tRNAs are available). Little is known about the effects of 3' –untranslated regions on gene expression besides the positive effects on RNA stability and translation of the poly(A) tail that is present on at least the vast majority of mRNAs.

Effects of Sequences Outside of the Expression Cassette

Even with thoughtful construction of expression cassettes, transgene expression, particularly in the progeny of a transgenic plant, is not always predictable. Sequences outside of the expression cassette can have profound influences on transgene expression. For other eukaryotes, such sequences can be located in the bacterial plasmid that harbors the expression unit.

Scaffold attachment regions

Scaffold attachment regions (SARs, also called matrix attached regions (MARs) are DNA segments that seem to organize eukaryotic chromosomes into domains of different activity. The flanking of transgenes by such regions reduces position effects. Transgene expression often is not related to the actual copy number of transgenes. This is explained by down-regulation or silencing of additional gene copies.

Optimized Transgene Expression Units

The basal expression characteristics depend on the respective promoter. Quantitative enhancement of expression independently of specificity is achieved by introduction of an efficiently spliced intron near the 5' –end of the transcribed region (particularly for cereal monocots) and the use of a polyadenylation signal from an efficiently expressed plant or plant virus gene. If the intended gene product is a protein, expression can be further promoted by an mRNA leader sequence (possibly derived from a plant virus) lacking secondary structure context, and a coding sequence devoid of clusters of rare codons, a start codon in a good sequence context, and a coding sequence devoid of clusters of rare codons. To reduce position effects, transgenes could be flanked by SARs, which may also inhibit gene silencing. It is as yet unknown whether sequences similar to the SP1 sites in some mammalian genes that inhibit methylation of adjacent genes also exist in plants and could be used to interfere with transgene methylation.

Expression Vectors

Most of the currently used expression vectors are simply a combination of strong promoters (mostly constitutive) with some polylinker sequences for simple cloning and a termination signal. A great number of such vectors for monocot and dicot transformation have been described. Vectors for monocot transformation often also contain introns to boost expression. Plasmids designed for *Agrobacterium*-mediated gene transfer require additional functions that allow maintenance in *Escherichia coli* and *Agrobacterium* (replicons and selectable markers) and transfer from *E. coli* and *Agrobacterium*, and T-DNA broder fragments for the transfer to the plant.

REPORTER GENE FOR TRANSIENT EXPRESSION

A reporter gene is added to gene of interest before its entry into the target cell. With the help of reporter genes, information on aspects of gene constructs such as the relative-level and tissue specificity of expression can be obtained.

Reporter genes are frequently introduced into genetically modified plants for identifying and tracking genetic modification as they tag the transformed cells. They find more use where the genetically modified plants can not be regenerated from single cell and direct selection is not feasible or effective.

The important features needed in reporter gene include its low background activity in plants and absence of any detrimental effect on metabolism. The gene product should have moderate stability *in vivo*, so that down regulation of gene expression as well as gene activation can be detected.

β-Glucuronidase Gene

The β-Glucuronidase (GUS) gene, encoded by the Uid A locus of *E. coli*, is most popular reporter gene used in plant transformation. The product of this gene i.e. enzyme β-glucuronidase breaks down glucuronides giving a coloured reaction. A blue coloration develops in the cells due to the reaction, thus expressing the gene. GUS activity can be quantified by excitation with light at 360 nm and monitoring fluorescence at 460 nm. Its presence can be detected *in situ*, inside the plant tissue, used either as thin section or in any other form. Several glucuronides which can be used as substrates, include β-nitrophenyl glucuronide (PNPG), 5-bromo, 4-chloro, 3-indolyl glucuronide (BCIG), naphthol As-BI glucuronide (NAG) and resorufin glucuronide (REG).

The major advantage of using this reporter gene lies in its assay, which is easy and sensitive. The expression of GUS gene fusions can be easily quantified by fluorometric assay, and histochemical analysis can be used to localize gene activity in transgenic tissues.

Luciferase Gene

This gene can be used as reporter gene and for transient gene expression. The activity of luciferase gene fusion can be assayed in transformed plants by their ability to grow in dark. The enzyme luciferase differs in bacteria and *Photinus pyralis* (firefly). While in bacteria, it consists of two peptide subunits coded by gene lux A and lux B, in case of firefly it consists of single polypeptide coded by a gene lux. The enzyme luciferase uses different substrates for luminiscence in these two systems. When the source of the gene is bacteria, the substrate used is an aldehyde (decanal) which is supplied exogenously with FMNH2. Alternatively when the source is firefly, luciferin (a heterocyclic carboxylic acid) is used as a substrate in the presence of ATP and oxygen. It generates oxyluciferin and yellow green fluorescence.

Chloramphenicol Acetyl Transferase Gene

This gene, isolated from *E. coli*, is primarily used as reporter gene and not as marker gene. As this gene is not present in mammals and higher plants, gene construct introduced in them can be detected by chloramphenicol acetyl transferase (CAT) assay. The enzyme uses acetyl Co A and, chloramphenicol (32P) as substrates and helps in the transfer of acetyl group to chloramphenicol. Presence of acetyl-chloramphenicol can be detected through autoradiography. CAT gene is also used for identification of a number of regulatory sequences.

Nopaline (nos) Synthase Gene

The disarmed Ti-plasmid obtained from *Agrobacterium tumefaciens* is used as a vector for gene transfer in dicot plants. The disarmed plasmid is without tumour causing region of TDNA. Hence it can not cause tumour. However, the border sequence and the nos gene are retained e.g. pGV3840. This plasmid is derived from the Ti-plasmid pTiC58. The nos gene codes for the enzyme nopaline synthase. Plant cells transformed using such a plasmid (pGV3840) will produce nopaline in their cells. The presence of nopaline can be identified by biochemical analysis. A positive result will show that cells are transformed.

Neomycin Phosphotransferase Gene (NPT II)

This gene is used both as selectable marker gene as well as reporter gene. This is the first marker to be used and is isolated from *E. coli.* This gene has been successfully used in many systems including mammalian cells and fungi. In solanaceous plants such as petunia, tobacco and tomato, it provides an excellent, unambiguous selectable marker. For assay of the gene product (i.e. neomycin phosphotransferase), this enzyme is first fractionated using non-denaturing polyacrylamide gel electrophoresis (PAGE). Since the enzyme detoxifies kanamycin by phosphorylation, radioactively labelled ATP (^{32}P) is also used with it and incubated at 35°C. The phosphorylation leading to incorporation of ^{32}P in kanamycin can be detected by autoradiography. Another method of enzyme assay is much easier. Dot blots of cell extracts are obtained on filter and this filter is incubated with substrate and finally autoradiographed to detect NPT II activity.

Anthocyanins

It utilises the C1, B and R genes, which code for trans-acting factors that regulate the anthocyanin biosynthetic pathway in maize seeds. The introduction of these regulatory genes into plants by microprojectile bombardment induces cell-autonomous pigmentations in non-seed tissues.

Dihydrofolate Reductase Gene

Another selectable marker is the mouse gene for dihydrofolate reductase. It encodes an altered protein that does not bind to methotrexate, a potent inhibitor of this essential enzyme. Petunia cells that contain and express this gene are able to grow on concentrations of methotrexate that completely inhibit the control cells.

For expression of these reporter gene, suitable promoters are added adjacent to the former.

PROMOTER ELEMENTS

One of the best studied plant promoters is the 35S promoters. This viral promoter produces the cauliflower mosaic virus 35S genomic RNA. In the early experiment approximately 1000 bp of promoter DNA was fused to various reporter genes. Analyses of protoplasts, transformed calli and transgenic plants demonstrated that the reporter genes with promoters were always expressed at high levels and were also insensitive to various endogenous and environmental cues (such as hormones, heat shock or light). The so called constitutive expression made the 35S promoters popular in dicot species. In monocot cells, the expression levels with the 35S promoter are 10 to 100 times lower than in dicot. A variety of promoters and other methods have been used to increase the level of expression of genes introduced into monocot cells. The incorporation of an intron into the transcriptional unit of the foreign gene increases gene expression in transformed monocot cells. The combination of the first intron of maize alcohol dehydrogenase 1 gene (Adh 1) and the CaMV 35S promoter have been used to control expression of the bar gene in transformed maize, oat and wheat. An alternative strategy is to include a modified maize Adhi sequence in the Emu promoter. This has been used in expression of neomycin phosphotransferase (npt II) gene in transformed rice and sugarcane. Finally, efforts have been made to isolate monocot promoters that show high level constitutive activity. The promoters of the genes for rice actin (Acti) and maize ubiquitin (Ubii) have been used to express the bar gene in transgenic wheat, barely and rice.

The direct transformation systems are most suitable for transient expression studies at present. However, the most successful and widely used system for stable transformation of plant is that based upon the Ti plasmid of *Agrobacterium.* In this method, the tissues are inoculated with *Agrobacterium* carrying the requisite vector having gene of interest. The next step is to select the transformed cells or tissues. This is facilitated by the presence of selectable marker genes available in vector.

MARKER GENE

A good selectable marker gene is one that provides resistance to a drug, antibiotic, or other agent that arrests the growth of normal plant cells (plant cells not harboring the marker gene). The selective agent should inhibit the growth of plant cells or kill them slowly. Agents that kill cells rapidly often result in the release of phenolic compounds and other substances (from the dead cells) that are toxic to the growth of the remaining, otherwise resistant cells. Various markers have been used for selection of transformed cells

Three marker genes among them are used more extensively, providing resistance to the antibiotics:

1. Kanamycin (and the related aminoglycoside G418,which are more effective growth inhibitor in eukaryotes),
2. Hygromycin, and
3. Methotrexate

Transformed system (tissues/organ) is placed on regeneration medium containing an antibiotic like **kanamycin** and transformed plants can be obtained directly. Cells, which are not transformed, die due to kanamycin. The heterogenous explant like calli or non-transformed tissue may escape initial selection. The leaf disc instead, exhibits the regeneration from edges, which is in contact with selective agents and thus, is a better alternative.

INACTIVATION OF FOREIGN GENE EXPRESSION

There are a number of conditions that can interfere with gene expression in transgenic plants. These include:

1. Poor recognition of promoter elements and differences in transcription efficiency between the source of the promoter and the transformed material.
2. Inefficient mRNA termination, polyadenylation, processing and/or stability.
3. Inefficient translation due to low rates of translation initiation and/or inappropriate codon usage.
4. Poor design of expression vectors, which might contain multiple repeated regions or multiple genes with opposite orientations leading to transcriptional interference.
5. Antisense-mediated gene inactivation.
6. Gene silencing by homologous sequence.

TRANSGENE INACTIVATION AND METHYLATION

Multiple copies of the transgene also suppress their expression. There is reverse correlation between gene copy number and expression level. Low expression is correlated with increased methylation of the transfer DNA. It is well established that methylated DNA is not or poorly transcribed. The most common DNA modification in plant cells is cytosine methylation at CG dinucleotides and CNG (were N can be any base) trinucleotides. Methylcytosine can interfere with protein-DNA interactions. This phenomenon has been observed with maize dihydroflavonol-4-reductase (DFR) gene in Petunia and with GUS transgene in tobacco.

DOUBLE TRANSFORMATIONS AND GENE SILENCING

Introduction of a second T-DNA construct into tobacco already containing T-DNA also lead to the suppression of genes located on T-DNA first. Presence of second T-DNA leads to methylation of the genes of T-DNA first, thus resulting in their suppression. If T-DNA second is crossed out, the T-DNA first genes become demethylated and reactivated. It has been suggested that homology between the DNA sequences of the two T-DNAs, even though they are not transcribed, triggers the silencing.

POST-TRANSCRIPTIONAL GENE SUPPRESSION

Suppression of β-1-3-glucanase and chitinase transgene expression is only observed in plants that are homozygous for the transgene. The β-1-3-glucanase transgene suppression is the result of a post-transcriptional process and is observed only in homozygous and in haploid plants.

A number of issues still need to be addressed before transformation technology can be exploited fully in the agronomic improvement of plants. These include those that act to regulate the efficiency of transcript termination, transcript stability, post-transcriptional modification and translation efficiency. An increased understanding of these regulatory components in plants will prove profitable for achieving the regulated expression of foreign genes in transgenic plants.

6

Gene Silencing and Antisense Technology

The instability of expression of the transgenes or even the endogenous genes.

Gene silencing in plants has provided interesting but intriguing insight and valuable information about gene regulation, especially when challenged by the alien interference. Earlier gene silencing was considered as a disappointment, now it turned as a new, academically interesting and commercially viable sector of genetic engineering.

A considerable variation have been found in the expression of transgenes in individual transformants which is not due to copy number (Stam et al., 1997). These transgenes are when inserted into a plant, they either promote the expression or suppression of homlogous genes. These means that transgenes can inactivate expression of homologous genes. The suppression of homlogous genes is known as gene silencing. This phenomenon has been reported in various plant species, transformed with various genes and is a form of epigenetic modification. The inactive state is usually stable and in some cases is meiotically stable, but it is not permanent. The silent genes can undergo spontaneous as well as developmentally regulated reactivation by a process. It is called **resetting**.

The use of genetic engineering in plant breeding is a routine technique for crop improved around the world, specially for development of transgenic plants. This technique has developed many problems, as the introduced genes (transgenes) though properly introduced genes (transgenes) though properly integrated, do not always express fully. The frequently observed instability of transgene expression, commonly called **gene silencing**. It is not restricted only to the transgenes, as even the endogenous genes are subjected to this suppression after transfer of a homologous transgene. Gene silencing is the setback to the idea of commercialization of transgenic plants, and this was the challenge for molecular plant breeders to unravel the cause (s) of gene silencing.

The expression of the desired phenotype in which the alien gene has been introduced depends on various factors, the important factors are:

1. The transgene needs to be integrated at a site where as transcription is not suppressed by **cis-effect** of the neighboring sequences.
2. Its inaccessibility to the transcription and processing machinery in the packed chromatin of the interphase nucleus.
3. To ensure expression of the transgene, its transcripts have to be properly processed and transported to the cytoplasm where these need to be precisely translated and the products suitably modified for ultimate expression.
4. A minor alteration at any of these steps may lead to the absence of the expected characteristic or development of an unusual phenotype or inheritance.

REASONS FOR GENE SILENCING

The gene silencing in plants is a genetic control mechanism implicated in virus resistance (Cover et al., 1997; Ratoliff et al., 1997), genome maintenance and developmental control. This concept was given by Vionnet and Baulcombe (1997). The signal molecule is likely to be nucleic acid. Probably, signal molecule could mediate sequence specific gene silencing by forming a base-paired or triple helical structure with the target RNA. A nucleic acid could move in plant using the channels involved in virus or viroid movement. Finally Hamilton and Baulcambe (1990) discovered that post-transcriptional gene silencing in plants is a nucleotide sequence specific defence mechanism that can target both cellular and viral m-RNAs. They analyzed three types of transgene-induced PTGS and one example of virus-induced PTGS in plants.

In each case, antisense RNA complementary to the targeted mRNA was detected. These RNA molecules were of uniform length, estimated at 25 nucleotides, and their accumulation required either transgene sense transcription or RNA virus replication. They suggested that the 25 nucleotide antisense RNA is likely synthesized from an RNA template and may represent the specificity determinant of PTGS.

To account for the sequence specificity and posttranscriptional nature of PTGS, is has been proposed that antisense RNA forms a duplex with the target RNA, thereby promoting its degradation or interfering with its translation. If these hypothetical antisense RNA molecules are of a similar size to typic mRNA, they would have been readily detected by routine RNA analyses. However, there have been no reports of such antisense RNA that is detected exclusively in plants or animals exhibiting PTGS. Nevertheless, PTGS-specific antisense RNA may exist, but may be too short for easy detection. Hamilton and Baulcambe (1999) carried out analyses specifically to detect low molecular weight antisense RNA in four classes of PTGS in plants. They used five tomato lives, each transformed with a tomato 1-aminocyclopropane-1-carboxylate oxidase (ACO) cDNA sequence placed down stream of the Cauliflower mosaic virus 35S promoter. Two lines exhibited PTGS of the endogenous ACO mRNA. Low molecular weight nucleic acids were purified from the five lines, were separated by denaturing polyacrylamide gel electrophoresis, blotted and hybridized to ACO sense (Antisense-specific) RNA probe.

They found that a discrete, ACO antisense RNA of 25 nucleotides was present in both PTGS lines but absent from the nonsilencing lines. 25-nucleotide ACO RNA of sense polarity and at the same abundance as the 25-nucleotide ACO antisense RNA was also present only in the PTGS lines.

Thus, Hamilton and Baulcombe (1999) reported 25-nucleotide antisense RNA, complementary to targeted mRNAs, accumulates in 4-types of PTGS. They detected 25-nucleotide RNA in other examples of PTGS, and never detected 25-nucleotide RNA in the absence of PTGS. This correlation and the properties of 25-nucleotide RNA were consistent with a direct role for 25-nucleotide RNA in PTGS induced by transgenes or viruses. The twenty five nucleotide RNA species also served as molecular marker for PTGS. Their presence could be used to confirm their examples of transgene – a virus induced PTGS and perhaps also to identify endogenous genes that are targeted by PTGS in nontransgenic plants. They also suggested that 25-nucleotide antisense RNA species are not degradation products of target RNA because they have antisense polarity. A more likely source of these RNAs was the transcription of an RNA template. This was consistent with the presence of the 25-nucleotide PVX-infected cells that did not contain a DNA template.

An RNA-dependent RNA polymerase, as acquired by this model, is required for PTGS in *Neurospora crassa*. The precise role of 25-nucleotide RNA in PTGS remains to be determined. However, it has been suggested that they are long enough to convey sequence specificity yet small enough to move through are components of the systemic signal and specificity determinants of PTGS.

TYPES OF GENE SILENCING

Homology-Dependent Gene Silencing

Homology-dependent gene silencing occurs very frequently when a number of copies of the transgene are integrated at the some locus or in different positions within the genome. This is described as the homology detection mechanism of plants which leads to methylation of the multiple homologous copies. Though the mechanism of this process has not been completely understood, however, the available evidences suggest that methylation is mediated by transient pairing of the homologous sequences. This occurs more efficiently, if multiple copies are integrated as concatamers at the same locus as inverted or direct repeats. The inactivation of transgenes after crossing with the lines carrying *hypermethylated homologous sequences* may be one of the reasons for the same pairing mechanism. Not all genes are susceptible to this type of trans-inactivation, reflecting that the capability of detecting the homology at different chromosomal locations varies. The telomeric region shown to be favorable site for interaction with the homologous sequences.

In transgenic tobacco plants, homology-dependent gene silencing has been observed. The expression of *one T-DNA* construct in transgenic plants was suppressed when a *second T-DNA* construct was introduced. Both these T-DNA constructs had the same nopaline synthase (NOS) promoter of *Agrobacterium tumefaciens*. This trans-effect was because of the promoter homology and is known as *transient pairing*. It results in the transfer of methylation pattern from the silent loci to the active ones. Similar effect has also been obtained, if the two constructs are brought together by sexual crosses, and the degree of silencing depends on the copy number. This type of gene silencing is similar to the phenomenon of paramutation as found in maize. The DNA-methylation results in transgene silencing which appears due to DNA-RNA and RNA-RNA interactions. Evidence of RNA mediated DNA methylation has been observed in tobacco and potato.

Position Effects

The position of integration of transgene may influence subsequent gene expression. The gene integrating in the hypermethylated inactive regions of chromatin sometimes get methylated, thus, remain silence. The expression of transgenic increases if it is integrated into an actively transcribed region.

Environmental Stress

Environmental stress may cause gene silencing. The silencing effect of the environmental stress can be duplicated in tissue cultures by addition of propionic acid or butyric acid. These acids induce hypermethylation. In transgenic petunias A1 gene of maize conferring a brick red color to the flowers. When these plants were grown in glass house, they produced exclusively brick red flower irrespective of the growing season. On the other hand, when these were grown under field conditions, they did not do so. The plants which had produced brick red flowers during the early season developed weakly colored ones later. The change in color was found to be associated with methylation of the transgene.

The de novo methylation due to environmental stress may be due to sequence composition of the transgene which happens to be deviated from the adjacent regions. It may be a **natural defense mechanism** of the plants against foreign DNA followed by methylated and thus silenced. Based on AT contents, a common isochore structure has been deduced for plant genes. The AT contents of bacterial, monocot and dicot genes differ and therefore these have different isochore compositions.

Antisense Gene Silencing

Antisense gene silencing involves the inversion of the normal coding regions between the promoter and terminator sequences. As a result, rather than the sense strand, antisense strand is transcribed. The RNA so-synthesized binds to the RNA transcribed from the normal gene. The resultant RNA-RNA complex cannot be translated into a functional protein. As a result gene expression is switched off. The **RNA duplex** is degraded by the normal cellular mechanisms.

Endogenous Genes Silencing by Co-suppression

Endogenous genes are present in a cell and are prone to gene silencing. This takes place if transgene happens to be homologous to the endogenous gene, and is thus capable of transcribing mRNA of the same strandedness as the transcript of the endogenous gene. The effect has been termed as **co-suppression** which requires co-ordinate transcription of both the endogenous and the introduced genes.

This kind of gene silencing has been studied in petunias. It appears due to the expression of chalcone synthase (CHS) gene. **Chalcone synthase** is a key enzyme in flavonoid biosynthesis. It is responsible for color of the flower. To have darker flowers, the gene are re-introduced in a flowering line of *Petunia hybrida* which produces deep violet flowers. However, instead of having the desired effect due to the expected over-expression of the gene, the progeny produced white or variegated flowers. The resultant **aberrant phenotype** may be due to inhibition in expression of the gene. The co-suppression may result because of the reintroduction of an endogenous gene leading to a significantly reduced expression or total switching off. Therefore, if development controls are introduced the silencing can be avoided.

The mechanism of co-suppression is still an enigma. However, it has been suggested that it is triggered by an ectopic interaction of the homologous genes which results in the production of aberrant RNA (aRNA), in addition to the normal mRNA. The aRNA molecules are recognized by the cell and sequence-specific degradation is affected. This results in destruction of mRNA as well.

Types of gene silencing

GENE SILENCING AND NUCLEIC ACID PAIRING	**(1) Cis inactivation**	It involves inactivation by methylation of inverted or direct repeats on the same DNA molecule and is usually encountered in direct gene transfer methods.
	(2) Trans-inactivation	It involves sequences on separate DNA molecules which may or may not be allelic. This type of interaction involves coordinated silencing of a transgene or homologous endogenous gene or two homologous transgenes and may be either reciprocal (cosuppression or sense or non-reciprocial in which only one allele or locus is changed (dominant / epistatic trans-inactivation).
GENE SILENCING AND TRANSCRIPTIONAL ACTIVITY	**(A) Transcirptional Gene Silencings (TGS)**	TGS is associated with heavily methylated and inactive promoter sequences
	(1) Chromosomal environment of the transgene	When a transgene integrates into a chromosomal region, that is heavily methylated and / or repetitive, it is silenced. This indicates that the character of the flanking DNA, methylated or heterochromatin, is imposed upon the inserted transgenes. A transgene integrated into

	hypomethylated DNA can also be transcriptionally inactivated
(2) Repeat induced DNA methylation	Transgenes of T-DNAs which are inserted as a DR (Direct repeat – T-DNA arrangement 'heat to tail') or an IR (Inverted repeat – head to head or tail to tail insertion) have a tendency to become inactivated. This is frequently associated with DNA methylation.
(3) DNA – DNA interactions and gene silencing in trans	A transgene locus that is silenced can silence homologous transgenes at ectopic loci. When the silencing locus is methylated, the target locus also become methylated. In the case of homologous promoters, this transfer of methylation can lead to transcriptional inactivation.
(B) Post transcriptional Gene Silencing (PTGS)	The PTGS mechanism appears to act on any RNA, i.e., homologous to the transgene that has activated it, thus from transgenes endogenous genes and from transiently expressed genes. Also viral RNA are attacked when transgenes are used that contain viral sequences. The proportion of transformants in which transgenes and / or homologous host genes are silent varies, but can be very high.

POST-TRANSCRIPTIONAL GENE SILENCING (PTGS)

In PTGS, promoters are active and the genes transcribed but m-RNA fails to accumulate. The transgene induced PTGS mechanism affects expression of the transgenes and of endogenous genes with which they share a considerable degree of sequence identity. The latter case is also known as co-suppression because the endogenous genes and the transgenes are silenced. Co-suppression was first observed with genes involved in flower pigmentation (Napoli, Lemieux and Jorgensen, 1990; Van der Krol et al., 1990) and in tomato fruit ripening (Smith et al., 1990). Silencing of the

Chalcone synthase (Chs) genes in Petunia flowers occurs post transcriptinoally. (Van Blokland et al., 1994; Metzlaff, O'Dell and Flavell, 1996).

Characteristics of PTGS

The reasons for this are not known exactly, but they seem to be related to transgene expression levels and the organization of the T-DNA inserts. Moreover silencing does not always occur in every cell of a particular tissue. In anthocyanin gene silencing, it is visible due to variations in colour patterns (Vander Krol et al., 1990; Van Blotland et al., 1994; Jorgensen, 1995, 1996). In some systems, PTGS is developmentally regulated. Silencing of uid A, Chitinase Kunz et al., 1996) and of β-1, 3 glucanase transgenes in tobacco (De Carvalho Niebel et al., 1995) does not occur in young seedlings but starts between 2 and 6 weeks after germination, depending on the expression level of the transgenes.

Possible Mechanisms of Post-transcriptional Gene Silencing

There is no universal of gene silencing, because of the complex and intricate nature of the expression pathway. The precise molecular mechanisms of gene silencing is still needed. The possible role of PTGS and its identification can be observed by the redu·tion in accumulation of specific transcripts. This reduction may result from RNA instability (from activation of a sequence specific RNA degradation process). However, the exact reason is still unknown.

Model of transgene expression level (RNA threshold level)

This model explains that the transgenes produce too much m-RNA that level exceeds a threshold thereby triggering a mechanism that specifically removes all homologous RNA's irrespective of their source this RNA threshold model gained support from observations whereby PTGS was correlated with highly expressed transgenes conferring viral resistance (Lindbo et al., 1993; Smith et al., 1994; Goodwin et al., 1996).

Plants that contain highly transcribed transgenes did not accumulate the transgenic mRNA's. They remain resistant to infection by an homologous RNA virus (Smith et al., 1994; Goodwin et al., 1996). Plants, which normally contain lower levels accumulated transgenic RNA remain sensitive to a viral infection. Sometimes a virus infection is needed to induce the PTGS mechanism and resistance. This result in line with reaching the hypothetical RNA threshold level (Lindbo et al., 1993). Studies by Metzlaff et al., (1996) of Chs cosuppression in *Petunia* transformant suggests that silencing requires active transgenes.

PTGS and Methylated Transgenes

The silencing may appear at both transcriptional and post-transcriptional levels. It is often accompanied by an increase in DNA methylation, especially in the former case. PTGS is frequently associated with methylated transgenes (Hobbs et al., 1990, Ingelbrecht et al., 1994; Smith et al., 1994; English et al., 1996). Also Smith et al., (1994) noticed that the viral transgenes of viral resistant plants were methylated more than those of sensitive plants. Despite this methylation, the promoters were active and the genes transcribed.

Methylation of an increased number of cytosine residues resulting in the formation of 5-methylcytosine moieties is a common feature of DNA found associated with transcriptionally inactive regions of chromation. The inactivation of transgenes is also due to increase in DNA methylation. Therefore, the ultimate effect (gene silencing) is preceded by DNA methylaiton, however, the causes leading to this event vary.

The direct evidences for a role of DNA methylation in PTGS was observed by English et al., (1996). The PTGS *ed uid A* genes in tobacco plants were mentioned by Hobbs et al., (1990, 1993). These genes were more methylated near the 3' end of the gene were which included the polyadenylation region. This 3' region was required to prevent infection by the PVX *uid A* chimeric viruses, as if this part of the RNA is the target for the degradation of RNA. These findings show a link between DNA methylation and the RNA based silencing mechanism.

RNA directed DNA methylation (RdDM)

In tobacco transformants, it was found that transgenes composed of 3 or 4 tandemly arranged viroid c-DNAs in sense or antisense orientation and driven by the CaMV-35S promoter were methylated. This methylation was caused by the viroid RNA's. A two copy and a Viroid cDNA construct which were unmethylated, became methylation had taken place, indicate that the viroid RNA is responsible for the methylation of the homologous transgene DNA, rather than the transgene structure or intergration site. The process of its occurrence is unknown, but DNA-RNA hybrids are formed (Wassenegger et al., 1994).

Elimination of homologous transcripts

If the PTGS mechanism is induced according the RNA threshold model, it shows that cells are able to measure levels of specific RNA's. In normal cells, if such activity functions monitoring RNA levels from endogenous genes, one would expect that it also establishes and maintains a certain steady state, RNA levels in transgenic cells. But the vast majority of the homologous RNAs are degraded. To speculate about the nature of this RNA degradation activity, Dougherty's group postulated a role for complementary RNAs (cRNA) synthesized by the plant encoded RNA dependent RNA polymerase (RdRP) in the cytoplasm (Lindbo et al., 1993; Dougherty's and Parks, 1995). RdRP from tomato was recently characterized (Schiebel et al., 1993a) and it was known that in vitro; this enzyme synthesizes small RNAs from RNA templates (Schiebel et al., 1993b). It is hypothesized that in vitro these c-RNAs hybridise to the complimentary m-RNAs which are then degraded by double stranded RNA specific RNAases. The appeal of the RdRP-cRNA model is that it explains the strong sequence specificity of PTGS.

PTGS of Endogenous genes and the presence of specific transgene loci

PTGS is often associated with multi copy T-DNA loci (English et al., 1996). How the organization of a transgene locus affects PTGS to explain this there are two possibilities.

(a) A multicopy locus is prone to deliver the hypothetical aberrant RNA assumed to trigger the cytosolic RNA degradation machinery directly. This requires transcription of the transgenes. The aberrant nature of the RNAs might originate from read through transcription, abnormal RNA processing or transcription of a methylated template (Smith et al., 1994; English et al., 1996). When the RNA degradation machinery is activated, it eliminates all homologous RNAs.

(b) The second possibility applies only to cases where the genome contains an endogenous homologoue. Given that barely transcribed Chs sense transgene and promoterless transgenes are able to confer strong suppression of endogenous Chs genes. (Van Blokland et al., 1994) a transgenes derived RNA plays a key role in this PTGS process in line with the RdRP-cRNA model (Dougherty and Parks, 1995). The aberrant RNAs might be derived from the endogenous Chs genes. One possibility is that it occurs as a result of ectopic DNA pairing (Jorgensen, 1992) between the transgene locus and the endogenous gene (s). Based on the detection of elevated levels of unspliced Chs transcripts in nuclei containing silenced chs genes, this pairing may have changed features of the endogenous gene leading to an unpaired processing and / or transport of the RNAs. These transcripts may be intrinsically unstable and rapidly degraded, or may act as aberrant RNA causing the degradation of other homologous RNAs.

Not all transgene loci, may be able to pair ectopically with an endogenous gene.

GENE SILENCING IN CROP PLANTS

Tomato

The polygalacturanase (PG) gene is involved in cell wall degradation during normal ripening process, e.g., tomato. The tomatoes produced by employing antisense or truncated sense PG gene, have longer shelf life. They all can be harvested at one time. They are more resistant to post-harvest fungal infections. They are less prone to damage during transportation. The engineered tomatoes contain more pectin and thus have higher viscosity and solid contents. The higher pectin content is retained because of the considerably decreased polygalacturanase activity which is also delivered to play some role in depolymerization of pectins during ripening process.

Brassica

Introducing desired genes for gene silencing in *Brassica* is also found because it increases the stearate acid in seeds of *Brassica rapa* and *B. napus* by antisensing the seed-specific steroylacyl carrier protein desaturase gene.

Rice

In rice, by employing antisense approach, reduction of some allergenic proteins can be obtained. Production of rice variety, better suited for brewing can be obtained by antisensing the gene responsible for synthesis of glutelins (major storage protein). Its higher content gives an undersirable *rough taste* and shorter shelf-life to the wine.

Development of Viral Resistance in Plants

The resistance against virus in many plants has been developed by transforming them with a part of the viral genome. As explained earlier, replication of the virus in such transformed plants is prevented because of the co-suppression effect.

Some Examples of Gene Silencing

Chitinase Gene Silencing in Tobacco

Class I chitinases are vacuolar enzymes implicated in the defense reaction of plants against microbial pathogens reviews see Broglie and Broglie, 1994;

Kombrink and Samsoich 1995; Meins et al., 1994). The tobacco enzymes show complex regulation at the m-RNA level.

Nicotiana sylvestris plants transformed with a tobacco Class I chitinase CHN 48 gene regulated by CaMV 35S RNA expression signals frequently showed silencing (Hart et al., 1992). Roughly, 20-60% of homozygous, sibling plants exhibited a silent phenotype, leaves accumulated less chitinase than comparable leaves of untransformed plants; levels of both the approx. 34 KD a transgene encoded chitinase A and approx. 32 KD a host encoded chitinase and their m-RNAs were greatly reduced and the normal regulation of the host chitinase was lost. The remaining sibling plants exhibited a high chitinase phenotype; leaves accumulated very high concentrations of transgenes encoded chitinase A upto 1 mg g^{-1} FW and regulation of the homologous host encoded chitinase was normal.

Inactivation in chitinase gene silencing appears to be at the post transcriptional level, results from the interaction of homologous transcribed sequences and depend on transgene copy number.

Effect of Transgene dose on silencing

A chimeric tobacco CHN 48 gene was introduced into N. sylvestris by leaf disc transformation with the Ti plasmid binary vector PSCH-12. Most of the experiments were performed with plants homozygous for the transgene derived from the transformant SSC 2.3. This transformant has 2-3 copies of the transgene inserted at a single locus (hart et al., 1992). Untransformed wild type N. sylvestris or the homozygous line SCI B_2 transformed with the empty vector, PCI B 200 were used as controls. Plants were raised from seed in petridishes under conditions known to promote a high incidence of chitinase gene silencing (Hart et al., 1992). Leaves of these plants were then screened for silencing by immunoblotting with antibodies that detect both tobacco and *N. sylvestris* Class I chitinases. Chitinase transformants were judged to express the silent phenotype if the chitinase content was the same or lower than comparable tissues of untransformed or empty vector transformed plants assayed in the same experiment.

To find out whether the incidence of silencing depends on the number of transgenic loci, crossing between two independent homozygous transformants SSC 2.3 and SSC 2.4, with each other with wild type plants is done.

CROSS	% SILENT
SSC 2.3 X	62.3
SSC 2.3 X WT	0.0
WT X SSC 2.3	0.0

CROSS	% SILENT
SSC 2.4 X	76.2
SSC 2.4 X WT	0.0
WT X SSC 2.4	0.0
SSC 2.3 X SSC 2.4	74.6
SSC 2.4 X SSC 2.3	60.5

Populations of homozygous SSC 2.3 and SSC 2.4 plants showed a high incidence of silencing, silencing was not detected in the progeny of reciprocal outcrosses of these lines to wild type plants. There was high incidence of gene silencing in F1 hybrids of SSC 2.3 and SSC 2.4 comparable to that of homozygous parents. These results indicate that silencing depends on the number of transgene loci and not on allelic interactions at the transgene locus. Non-allelic copies of the same transgene on different chromosomes can interact to give a silent phenotype.

Silencing depends on similarity of transcribed sequences

Tobacco transgenes and homologous host chitinase genes with different promoters can be coordinately inactivated suggesting that silencing depends on similarity of transcribed sequence (Hart et al., 1992). This was confirmed by comparing the incidence of silencing in independent transformants obtained with chitinase and β-1,3 gluconase A (GLA) gene was chosen because GLA and CHN 48 show very similar developmental, pathogenesis related and hormonal regulation at the RNA level, but have unrelated DNA sequences. Chitinase levels were not reduced in β-1,3 glucanase silencing and β-1,3 glucanase levels were not reduced in chitinase transformants showing chitinase silencing. These results which were also confirmed by immunoblot and RNA blot hybridization analyses, indicate that β-1,3 glucanase and chitinase genes undergo independent silencing in response to homologous tobacco transgenes. So silencing in this experimental system depends on similarity of transcribed sequences rather than similarity of promoter sequences.

Class II chitinase host genes are not silenced by a Class I transgenes

The major difference between these families is that the Class II genes encode chitinases lacking the lectin domain, spacer and C-terminal extension of Class I enzymes.

The Class II isoforms are not usually expressed in healthy leaves, but are induced as part of the hypersensitive response to pathogens.

Silencing occurs during seedling development

To establish when silencing occurs, the chitinase content of homozygous SSC 2.3 and empty vector (SCI B2) seedlings raised under the same condition in the same experiment. Silent plantlets first appeared in SSC 2.3 populations at the 6 leaf stage, 45 days after the start of germination. By the 8-10 leaf stage 61 days, 98% of the seedlings were of silent type.

Stability of the silent state during vegetative growth

In an earlier preliminary experiment, two of five mature homozygous SSC 2.3 transformants exhibited a discontinuous gradient of chitinase expression in which leaves / below leaf 23 (counting from the bottom of the plants) showed a high chitinase phenotype and leaves / above leaf 22 showed a silent phenotype. In the present study, a survey of large population of 217 SSC 2.3 plants were done and it was found out that about 20% showed variable silencing in different leaves.

Spatial patterns of silencing were also detected in the stem. Tissue prints were made of longitudinal and transverse stem sections and stained with antibody directed against Class I chitinases. In one representative flowering SSC 2.3 plant,. Stem tissue from the uppermost leaf node (leaf 36) dam to leaf node 7 showed a uniform high expressing phenotype, whereas stem tissue below leaf node 7 showed a uniformly silent phenotype.

Spatial patterns of silencing were found with 6 independent chitinase transformants and the β-1,3 glucanase transformant SSG 7.2. Variegation in silencing is not a unique feature of a single transformation event and occurs in both chitinase and β-1,3 glucanase transformants.

Persistent of the silent state in non-growing meristems was examined by measuring chitinase expression in lateral shoots. Mature plants were decapitated to activate lateral bud development and leaves on the resultant lateral shoots as well the subtending leaf on the main axis were then assayed for chitinase content. Of 36 high expressing and 21 silent leaves surveyed, leaves of lateral shoots always showed the chitinase phenotype of the subtending leaf.

GENETIC FACTORS INFLUENCING SILENCING

Comparison of transformants carrying transgenes with the same 35S expression signals and different transcribed sequences indicates that competence for chitinase gene silencing depends on similarity of transcribed sequences rather than the promoter region.

The full length tobacco CHN48 transgene, the truncated CHNA transgene and the homologous Class I chitinase genes of the host genes were silenced in hybrids of homozygous SSC 2.3 and SCM 4.5 transformants indicating that homology between the regions encoding the catalytic domain plus 7 amino acid long C-terminal extension of the genes is sufficient to trigger silencing. The tobacco transgene did not silence Class II chitinase genes of the host encoding a catalytic domain that had diverged by about 40% from the Class I genes. This suggests that more than 60% identity of sequences encoding the catalytic domain is needed for silencing.

Chitinase gene silencing was detected in homozygous but not in hemizygous monogenic transformants. Similar effects have reported for silencing of N. plumbaginifolia β-1,3 glucanase (gn 1) (De Carvalho et al., 1992). Nia (Dorlhac de Borne et al., 1994) and NiR (Vaucheret et al., 1995) transgenes in tobacco. Genetic crosses of different independent transformants with inserts at unlinked loci show that competence for chitinase gene silencing depends on transgene dose rather than on interaction of alleles at the same transgene locus.

Silencing of chitinase gene expression is likely to be a post transcriptional event

Although steady state levels of chitinase transcripts were greatly reduced in silent leaves, that of nascent transcripts produced by nuclei isolated from silent and high expressing tissues were similar. These results are consistent with the hypothesis that chitinase gene silencing in N. sylvestris is a post transcriptional event.

Occurrence of Silencing and Resetting

Once established, the silent can persist. When N. sylvestris plants are grown in the greenhouse under conditions that do not induce chitinase gene silencing (Hart et al., 1992). In some mature plans, all leaves tested showed the silent phenotype. This means that at least some silent leaves are derived from initials formed from the shoot meristem under non inductive conditions. The silent state usually persists in leaves derived from shoot meristems formed de novo from cultured leaf disks. These results strongly suggest that silent state or the potential to from leaves exhibiting the silent state can be maintained in growing shoot meristems.

In contrast to the triggering of silencing, which is stochastic, resetting occurs regularly. Seeds produced by silent plants and young seedlings germinated from these seeds invariably show the high chitinase phenotype. High levels of chitinase gene expression in developing seeds of silent flowers are first detected about 8-11 days after pollination suggesting that resetting occur at this time.

Chitinase gene silencing represents a less stable form which is not meiotically heritable shows pronounced developmental and environmental regulation. It appears to be post-transcriptional and probably involves interactions between transcribed sequences.

Sense Suppression of Pigment Genes in Petunia Flowers

Sense suppression is a post transcriptional phenomenon usually resulting in the co-suppression of homologous transgenes and endogens. Sense suppression is a non-linear inverse response that appears to be threshold dependent and so is able to amplify small, quantitative changes in gene dosage or transcription into large qualitative changes in phenotypic expression.

In the case of a flower pigment gene such as chalcone synthase (chs) the effect is a visible loss of flower pigments. However, rather than uniform reduction in pigmentation across the corolla, an altered spatial distribution of pigments produces a wide variety of striking patterns. These patterns range from simple to complex, ordered to disordered and stable to metastable. The white, pigmentless sectors that define the patterns are organized primarily according to petal morphology rather than clonal cell division patterns. A broad boundary zone in which the pigment intensity declines in a steep continuous gradient separates fully pigmented sectors from white sectors (Napoli et al., 1990; Jorgensen, 1995). To explain these patterns, it has been proposed that transcription factors responsible for Chs gee expressio vary in concentration at different positions in the petal in a maner controlled by fields of morphological information. Such spatial variation in transcription factor concentration could then produce a gradient of Chs transcription rate across the petal, reflecting its morphology. Wherever the gradient of Chs transcript concentration corss is the threshold for induction of sense suppression it will be amplified into a steep, inverse gradient of transcript turnover rate, which will be visible as a steep gradient of anthocyanin pigment.

It has been demonstrated that sense suppression phenotypes can be just as stable as phenotypes controlled by naturally occurring genes (Gutterson, 1994). The threshold dependence of sense suppression would seem to require that a sense transgene be expressed at levels reliably above the threshold genes. Anthocyanin pigmentation is well known to be a useful phenotype for gene expression studies, especially for investigations of biological patterns, because it is visible, cell autonomous and nonessential.

For study of epigenetic events, a herbaceous perennial like Petunia can be advantageous

(1) The growth habit of Petunia plants is easily manipulated by pruning to release multiple points of lateral branching.
(2) Petunias can be vegetatively propagated from stem segment cuttings.
(3) Petunias can be maintained in a continuously flowering state.
(4) Despite being a long lived perennial, Petunia can flower within 2 months of seed germination.

Napoli et al., (1990) reported substantial variations in the patterns of Chs co-suppression both within and among transgenotes and between them and their sexual progeny.

ANTISENSE TECHNOLOGY

Calgene created in 1994 the first commercial transgenic plant, a long shelf life tomato, by the suppression of polygalacturonase activity due to an antisense gene (Smith, 1988). However, this Flavr Savr tomato variety was removed from the trade 3 years later, because of its disease susceptibility and its lack of productivity. Later, other tomato varieties with long storage qualities were obtained by the utilization of an antisense RNA inhibition of ACC synthase or ACC oxydase, two ethylene precursors.

The antisense technique was also used to reduce the lignification of woody plants, by blocking the enzymes involved in the precursor of lignin biosynthesis. Another interesting application was the induction of white flowers in petunia and in different other ornamental plants by the suppression of chalcone synthase activity.

This technology involves the reduction in gene expression by using RNA containing complementary sequences to a given portion of double helix transcribed into RNA it is always the same strand for a given gene. If a cloned gene is engineered so that only the opposite DNA strand is transcribed, the result will be "anti-sense RNA" with a sequence complementary to the normal RNA transcripts. Antisense RNA when synthesized in large amount will often hybridize with sense RNA (made by normal genes) thereby inhibiting synthesis of corresponding protein.

Like sense genes, antisense genes are also known to elicit post-transcriptional gene silencing. Suppression of gene activity by antisense genes has been used for inhibiting viral replication (Bejarano and Lichtenstein, 1992). Techniques of tomato expressing antisense RNA against protein encoded by Cl gene showed resistance to Tomato yellow leaf curl virus (TYLCV) (Bendahmane and Groneborn.,1997)

Antisense genes, initially found in Prokaryotes play a significant role in regulation of diverse and complex phenomenon like plasmid replication, regulation of phage reproduction etc. (Green et al., 1986). They play an important role in regulation of transgenic expression as well.

Formation of Antisense mRNA

(1) Antisense mRNA can be produced accidentally if the transgene has been introduced in a transcriptionally active region where there is a promoter to initiate transcription from the coding strand of DNA.

(2) Antisense mRNA can be transcribed if the transgenes are present with converging promoters in the construct such that if transcription over runs the termination signal, small amounts of RNA would also be made from the other insert but in opposite direction (Grierson et al., 1991).

(3) Antisense m-RNA can also be produced by RNA dependent RNA polymerase.

Inhibition of gene expression by Antisense RNA

Antisense m-RNA can hamper gene expression in many ways:

It can block RNA production by inhibiting transcription ie Antisense transcripts act in the nucleus by either interacting with DNA template to generate a triple helix (DuvalValentin et al., 1992) or with nascent transcript and thus cause transcription termination.

It may also interact with the promoter region to repress transcription.

(1) Gene expression can also be adversely affected due to interference of antisense RNA with m-RNA processing and transport.

(2) Antisense RNA may disturb normal transport of m-RNA out of nucleus.

(3) Antisense RNA also disturbs the regular flow of transcripts by forming hybrids with their target m-RNA.

(4) In Prokaryotes antisense RNA complementary to the ribosome binding sites can inhibit translation (Green et al., 1986) while inhibition in Eukaryots depends on the ability of antisense m-RNA to reach the cytoplasm.

Plants from Antisense Technology

Fruits ripening can be inhibited by antisense RNA

During tomato ripening, two ACC synthase genes are expressed ie LE-ACC_2 & Le-ACC_4. Expression of antisense RNA derived from LE-ACC_2 results in almost complete inhibition of m-RNA accumulation. The antisense phenotype can be reversed by treatment with ethylene or propylene.

Pigmentation in flowering plants

Pigmentation in flowering plants can be altered by this method. Chalcone synthase is key enzyme in flavonoid synthesis. It has been shown that constitutive expression of an antisense Chalcone synhtase gene *t* in transgenic *Petunia* and Tobacco results in altered flower pigmentation. It was observed that there was reduction in levels of both m-RNA for the enzyme itself. The trait was stably inherited.

Production of male sterile plants

Antisense RNA technology has been used for production of male sterile plants of *Brassica.* It has great potential as it will have applied values and will help to understand fundamental mechanisms of gene expression and development of organisms.

7

Transgenic Plants and Photosynthetic Carbon Metabolism

Transgenic plants refer to those genetically modified plants in which functional foreign gene(s) have been inserted in their genome. Recent advances in plant molecular biology, tissue culture and recombinant DNA technology have opened this new avenue in transformation of higher plants. Thus, transgenic plants have a piece of alien DNA that are introduced by procedure other than sexual crossing and thus are genetically modified. Crops with transgenic plant are now common feature of the agricultural landscape (Dunwell, 1999). For improvement in crop yield genetically modified crops have been developed because carbon metabolism in photosynthetically active cells of higher plants is of central importance for the allocation of carbohydrates formed to tissues that are dependent on the supply of carbon and energy. As discussed earlier, many control mechanisms influence carbohydrate formation and partitioning into different pools.

Photosynthetic process that occur exclusively within the chloroplast as stated earlier can be divided into light-dependent and light-independent reactions. The light-dependent reactions convert electromagnetic energy into biologically convertible energy forms such as ATP and NADPH. These are used to reduce CO_2 in the reductive pentose phosphate cycle (RPP cycle; Calvin cycle) reactions, which are chemically light independent, but nevertheless can be light regulated.

MANIPULATION OF PHOTOSYNTHETIC CARBON ASSIMILATION IN TRANSGENIC PLANTS

Transgenic tobacco plants which express the gene for the small subunit of Rubisco in the antisense orientation, there by producing reduced level of the protein, were the first example to be reported with respect to the manipulation of the level of an enzyme involved in the Calvin cycle through genetic engineering. Recently, the manipulation of Rubisco activity through the reduction of rubisco activase content in tobacco plants has also been reported. Genetic manipulation of starch synthesis in transgenic plants has hitherto also been reported for potato, tomato, and rice (Müller-Röber et al., 1992; Stark et al., 1992).

Manipulation of Rubisco activity in Transgenic Tobacco Plants

Rubisco catalyses the fixation of CO_2, and makes up to 50 per cent of the soluble protein in leaves of higher plants. Nevertheless, it is though that Rubisco is rate limiting for photosynthesis. Rubisco has a low specific activity and very poor affinity for CO_2. It not only catalyses the carboxylation of RuBP, but 25-30 per cent of its activity at atmospheric CO_2 and O_2 concentrations is accounted for by the oxygenation of RuBP, a reaction leading to net CO_2 release, since one of the products, 2-phosphoglycolate, is oxidized in the photosynthetic carbon oxidation cycle in the process referred to as photorespiration. It is, therefore, necessary to optimize Rubisco with respect to its specific activity and affinity towards its gaseous substrates. Within Rubisco, many regions and residues have been found to be of catalytic significance. Site-directed mutagenesis makes it possible to exchange amino acid residues specifically within the protein. Earlier, Rubisco engineering was very difficult to increase its higher efficiency with respect to CO_2 fixation.

Studies with Tobacco

To solve this problem tobacco plant was used. Tobacco plants with reduced Rubisco content were analysed with respect to their photosynthetic activity, their growth, and their carbon allocation under different environmental conditions. Under an ambient light regime (310 μmol photons m-2 s-1) it was shown that Rubisco content can be decreased down to 50-60 per cent of wild-type levels before any effects on the photosynthetic activity of tobacco leaves are observed, clearly indicating that Rubisco is present in excess. It is now established that Rubisco is present in excess in wild-type tobacco plants, but its function as a pool for reduced nitrogen is still a matter of debate. Experimental evidences indicate that photosynthetic activity was not affected over a wide range of Rubisco reduction, therefore, neither the production of biomass was also not affected, nor was starch synthesis over the same range of inhibition. When Rubisco was reduced in a way that photosynthetic rates declined, lower biomass production and lowered starch synthesis in leaves were observed.

Rubisco activity is positively regulated through carbamylation by Rubisco activase. Furthermore, it is possible that the activase promotes the release of the inhibitor carboxy-arabinitol 1-phosphate from the enzyme in tobacco plants. Interestingly, lower Rubisco activation was compensated by higher Rubisco levels in the transgenic plants.

The tobacco plants do not possess strong sink organs that demand photoassimilates, such as seeds, root, or soil-borne tubers. Instead, breeding programmes have probably maximized the retention of assimilates within the leaves, since these are the organs of agricultural interest. Therefore, the results obtained with tobacco plant with respect to the limitation of photosynthesis by Rubisco are not necessarily true for other crop plants in which a higher demand for carbohydrates by sink organs is made, are consequently, a higher proportion of photoassimilates exported, out resulting in a higher rate of recycling of inorganic phosphate.

Further, such plants show higher photosynthetic activity, and it is not possible to tell whether or not rubisco is rate limiting.

Studies with Potato

In potato, the activity of the plastidic FBPase, a highly regulated enzyme involved in the regeneration of RuBP in the Calvin cycle, is reduced using the same strategy. Plants with as much as 40 per cent of wild-type FBPase activity show inhibition of photosynthesis when measured under maximal conditions. When grown under ambient conditions these plants did not show any reduction in tuber yield as compared to controls, indicating that the entire photosynthetic process (light and dark reactions) is not limiting for maximal tuber yields.

The plastidic FBPase antisense plants also serve as a good tool to investigate the rate limitation of photosynthesis under water stress, since it was shown that lowered amounts of Rubisco do not have any significant effects on this phenomenon. It is therefore possible that enzymes involved in the regeneration of RuBP are responsible for the decline of photosynthesis under water stress.

Manipulation of Triose Phosphate Translocator Activity in Transgenic Plants

It has been a matter of debate for a long time whether photosynthates are exported as triose phosphate or as hexoses from the chloroplast into the cytosol. Furthermore, it is an open question whether this export is rate limiting for photosynthėsis. Biochemical studies with spinach chloroplasts clearly indicate that triose phosphate export is the major route for carbon channeled into sucrose for further transport into sink organs, since other transport activities could not account for the sufficient translocation of metabolites across the chloroplast membrane during photosynthesis.

Glycolysis, sucrose biosynthesis, and sucrose transport

Photoassimilates enter the cytosol as triose phosphates or two major possibilities exist for the subsequent utilization. Triose phosphates can be used either in respiratory reactions or for sucrose biosynthesis. Several attempts have been made to analyse the influence and function of key enzymes involved in these pathways by creating transgenic plants. The interest behind these experiments is to understand the mechanism controlling partitioning and allocation of carbohydrates in higher plants. To increase crop productivity it would be useful to shift the direction of triose phosphate flow from glycolysis to sucrose synthesis, as the amount of sucrose present in the source leaf is a rate-limiting factor with respect to the capacity of a source leaf.

Attempts of manipulate the glycolytic pathway

After transport into the cytosol triose phosphate may enter the glycolytic pathway. This pathway ends in the cytosol at the level of pyruvate which subsequently enters the mitochondria for further metabolism, such as respiration amino acid or fatty acid synthesis. Pyruvate kinase (PK) catalyses the formation of pyruvate and ATP from phosphoenolpyruvate (PEP) and ADP.

It is known that there are multiple isozymes of PK in plastids (PK_p) and the cytosol (PK_p) and that there are complex regulatory mechanisms controlling their expression. In a recent study the potato PK_c was fused to a Rubisco transit peptide to target the protein to plastids in transgenic tobacco plants. Surprisingly, expression of both the resident and transgenic PKc was eliminated in some plants as a result of the phenomenon of co-suppression. Plant lacking any PKc grew and reproduced normally. Lack of PK would be fatal in other organisms and it is assumed that plants have some means of bypassing the reaction catalysed by PKc, perhaps through the action of PEP phosphatase or by the combined action of PEP carboxylase, malate dehydrogenase, and malic enzyme. A better understanding of the glycolytic pathway can be achieved through the modification of the levels of the various PK forms in transgenic plants and subsequent biochemical and physiological analysis.

Manipulation of sucrose biosynthesis

The major controlling steps of sucrose biosynthesis in higher plants are the interconversion of fructose 1,6-bisphosphate and fructose 6-phosphate, the formation of sucrose 6-phosphate from UDPG and fructose 6-phosphate, and the dephosphorylation of sucrose 6-phosphate yielding the final product, sucrose. Three enzymes are involved in catalyzing the first reaction: the forwards reaction is catalysed by fructose 1,6-bisphosphatase (FBPase), the backward reaction towards glycolysis is controlled by phosphofructokinase (PFK), and the third enzyme, pyrophosphate: fructose 6-phosphate-1-phosphotransferase (PEP), is able to catalyse the reaction in both directions. Each of these enzymes is allosterically regulated by intermediates of carbohydrate metabolism, furthermore FBPase and PEP are highly regulated by the regulatory molecule fructose 2,6-bisphosphate (F2,6P).

To date, two approaches using transgenic plants have been employed to manipulate the interconversion of F1,6P and F6P using transgenic plants. As the precise function significance of PEP is still unclear and controversial. The transformed potato plants with antisense constructs to the genes encoding the ∝- and β-subunits of PEP. In several transgenic plants PEP expression was decreased by up to 99 per cent. These plants did not show any phenotypical changes, reduced growth, or a decrease in tuber yield. The reduced PEP activity leads to a small reduction in starch and to a small increase in sucrose in potato tubers. These results were unexpected, because the essentially complete reduction of an enzyme that is highly regulated and expressed in most tissues was expected to influence plant carbohydrate metabolism significantly.

UDP-glucose (UDPG) is derived from glucose 1-phosphate by the action of the enzyme UDP-glucose pyrophosphorylase (UGPase) and represents a branchpoint in sugar metabolism: UDPG is needed for sucrose biosynthesis and as a precursor of cellulose. Transgenic plants with only 4 per cent of the wild-type UGPase activity showed no phenotypic changes, and sugar metabolism was not substantially affected (Zrenner et al., 1993). The fact that neither the suppression of PEP nor suppression of UGPase leads to major phenotypic effects could have two meanings. Either plants express certain enzymes in such an excess that even the low remaining activity is sufficient for normal development and growth, or these enzymes even when they are expressed in a more or less constitutive way are only needed under certain, as yet unknown, environmental or stress situations.

The reversible reaction catalysed by PEP may be shifted towards sucrose synthesis by removal of the phosphate group donor inorganic pyrophosphate (PP). In a recent study, *E. coli* inorganic pyrophosphatase (PPase) was introduced into the cytosol of transgentic tobacco and potato plants (Sonnewald, 1992). The expression of PPase in transgenic plants leads to a dramatic increase in the level of soluble sugars in source leaves. Although an accumulation of starch is observed in older leaves, the ratio between soluble sugars and starch is increased up to three times. This result could mean that expression of PPase leads to the expected alteration of carbon partitioning. The irreversible of F1, 6P and f6P is shifted towards F6P in the direction of sucrose synthesis. In addition, the formation of UDPG catalysed by the enzyme UGPase becomes irreversible due to removal of PPi and based on an increase in free Pi in the cytosol, the export of triose phosphate from the chloroplast is triggered. Growth and development of these plants are changed drastically. The plants are characterized by stunted growth and loss of chlorophyll. Potato plants showed an increase in tuber number, arguing for a reduction of sink capacity. As the PPase was expressed in all tissues several explanations are possible for the observed changes. Expression in sink tissues may prevent the utilization of UGPG which is produced by the cleavage of sucrose by the enzyme sucrose synthase. Additionally, PPase activity in companion cells may block phloem loading (Sonnewald, 1992).

A growing body of work indicates that sucrose phosphate synthase (SPS) regulates the synthesis of sucrose in the cytosol and, therefore, may play a major role with respect of the amount of photoassimilates available for export from source leaves (Stitt and Quick, 1989). The maximal activity of SPS is regulated in a light-dependent manner through phosphorylation or dephosphorlation, and furthermore through G6P and Pi (Huber and Huber, 1992). Different species show a variation with respect to light activation of SPS (Huber et al., 1989). Maize SPS activity, for example, is strongly down-regulated during the dark, but the enzyme does not respond to G6P or Pi concentrations in the light (Huber et al., 1989). The expression of the maize SPS in transgenic tomato plants caused a reduction of starch and increased the amount of sucrose in mature leaves (Worrell et al., 1991). This result clearly demonstrates that SPS is involved in the regulation of carbon partitioning in leaves.

Sucrose loading into the phloem

In photosynthetically active tissues a surplus of fixed carbon is produced, which is transported to sink tissues, e.g. roots, seeds, tubers, and sink leaves. Solutes are transported within the phloem from source to sink tissues. In this section we will concentrate on the transport of sucrose from mesophyll cells into the **sieve element/companion cell complex** (SECCC). Whether phloem loading is exclusively symplastic or if an apoplastic transport step is involved has been the subject of many discussions (Lucas and Madore, 1988; Bush, 1933). The distribution of plasmodesmata has provided evidence for symplastic transport, at least between mesophyll cells. In certain plant species, e.g. Laminaceae, a high symplastic connectivity between mesophyll cells and the SECCC (type 1 plants) may allow phloem loading via plasmodesmata, whereas in species like potato, sugar beet, and tobacco (type 2b plants), with comparatively low symplastic connectivity (van Bel et al., 1993), an apoplastic step is proposed. The apoplastic loading model predicts an energy-dependent transport step of sucrose across the SECCC plasma membrane, generating a sucrose gradient with a tenfold difference in concentration. Transport is driven by a proton gradient generated by plasmalemma H+-ATPase. In a variety of experiments a sucrose / proton symport activity was undoubtedly detected in plasma membrane vesicles and protoplasts of several plant species (Bush, 1993). There are two explanations for such a transport activity. First, the transport protein is part of a general sucrose retrieval system. Secondly, the protein is responsible for the active transport of sucrose into the SECCC. Recently, transgenic plants were generated which gave a deeper insight into the in vivo processes related to the questions concerning phloem loading.

Apoplastic expression of yeast invertase

Two groups have used the expression of a yeast invertase (Suc_2) in the apoplastic space in transgenic tobacco, potato, *Arabidopsis*, and tomato plants to get an insight into the sucrose loading process (Dickinson et al., 1991). In these studies significant phenotypic changes were reported. Transformed tobacco plants showed stunted growth, suppressed root formation, and the development of bleached and necrotic regions in older leaves. In addition, these leaves accumulate high levels of carbohydrates and starch. These carbodhydrates showed no diurnal turnover and the accumulation was accompanied by an inhibition of photosynthesis (Dickinson et al., 1991). Transgenic potato plants showed similar phenotypic changes, but only leaves near senescence showed bleached and necrotic region (Heineke et al., 1992). The observed decrease of photosynthesis is not caused by a feedback regulation of sucrose synthesis, but is accompanied by an increase in osmotic pressure in the leaf.

Reduction of the expression of sucrose transporter in transgenic potato plants

The transgenic plants show reduced enzyme activity due to the *antisense* technology allowing new insights into the mechanisms present in vivo. A prerequisite for this approach is the isolation of the gene to be suppressed. Recently sucrose transporter cDNA clones from spinach and potato were isolated by complementation of a yeast mutant (Riesmeier et al., 1993). The transporter proteins from both spinach (SoSUT) and potato (StSUT) showed characteristics consistent with those reported for the sucrose / proton symport in isolated plasma membrane vesicles. RNA *in situ* hybridization data show that the transporter gene is especially expressed in the minor veins of mature leaves (Riesmeier et al., 1993).

To further investigate the physiological function of the isolated cDNAs, potato plants were transformed with an *antisense construct* (Riesmeier et al., 1992). The antisense inhibition leads to drastic phenotypic effects. The growth of the plants is strongly retarded; mature leaves are curled, bleached, and accumulate anthocyanines at the rims, whereas sink leaves showed no phenotypic changes. Measurement of the export of soluble sugars through the petiole provides direct evidence that the flow of sucrose is reduced in the antisense plants. The reduced efflux strongly affects the supply of carbohydrates to sink organs, clearly visible as a reduced root system. No tubers are formed in the case of plants showing the highest reduction of StSut mRNA. The plants were analysed more intensively with respect of physiological changes. The soluble carbohydrate content in mature leaves increased up to twenty fold and the starch amount up to tenfold. The maximal rate of photosynthesis decreased to 50 per cent in strongly affected plants as compared to wild-type.

Starch Synthesis in Photosynthetic Tissues

It is regulated in a complex manner. Probably only part of the controlling mechanisms have been identified. Starch is produced in the chloroplasts, where part of the carbon fixed by the Calvin cycle is channeled along the pathway starting at fructose 6-phosphate (F6P). F6P is converted to ADP-glucose by the action of the enzyme phosphoglucoisomerase (PGI), phosphoglucomutase (PGM), and ADP-glucose pyrophosphorylase (AGPase). ADP-glucose serves as a direct precursor for starch production, which involves a complex array of enzymes including different starch synthases (soluble and granule bound isoforms), branching enzyme, and, more than likely, other unidentified propteins. During the day, i.e. when temporary storage of starch occurs, only a part of the recently fixed carbon is immediately available for regeneration of RuBP in the Calvin cycle or for conversion to sucrose in the cytosol. At night, degradation products are converted to sucrose, allowing transport of carbon metabolites to heterotrophic tissues. A precise control of leaf carbohydrate metabolism is needed to organize these processes efficiently and to ensure efficient growth of the plant.

ADP-glucose pyrophosphorylase

ADP-glucose pyrophosphorylase, a major enzyme controlling starch synthesis, has been demonstrated in many different plant species, including lower plants, monocots, and dicots. The enzyme is also present in bacteria, where ADP-glucose is the substrate for glycogen synthesis. Two long-known features of plant AGPase are the substrate for glycogen synthesis. Two long-known features of plant AGPase are the allosteric activation by 3-phosphoglycerate (3-PGA) and the inhibition by inorganic phosphate. In contrast, the enzyme from enteric bacteria is mainly regulated by fructose 1,6-bisphosphate (F1,6P; activator) and AMP (inhibitor) (Ghosh and Preiss, 1966; Plaxton and Preiss, 1987).

Following data underscores the importance of AGPase in starch synthesis:

1. AGPase is localized in chloroplasts and amyloplasts, i.e. in starch-storing organelles.
2. The in vitro, activity of AGPase play an important role in in vivo rate of starch accumulation.
3. Kinetic models led to the conclusion that 3-PGA and Pi, which are known to influence AGPase activity in vitro, play an important role in controlling starch synthesis in vivo.
4. A *Chlamydomonas reinhardtii* mutant with an unregulated AGPase (no activation by 3-PGA and no inhibition by Pi) produces only 5 per cent of starch seem in the wild type.
5. Mutats deficient for AGPase also have reduced levels of starch.
6. Transgenic plants in which AGPase activity was modulated showed altered levels of starch.

Plant AGPase has been described as a heterotetrameric protein composed of two different but similar polypeptides. The size of the two smaller subunits ranges from 50 to 55 kDa, the size of the larger subunit from 51 to 60 kDa, depending on the plant. At least ten different AGPase polypeptides from angiosperms, e.g. wheat, maize, rice, and potato, have been characterized at the DNA level (Anderson et al., 1989). Based on biochemical and DNA sequence data, the large subunit polypeptides appear to be more variable than the small subunit polypeptides exist in the potato tuber (Okita et al., 1990).

Transgenic plants with altered levels of AGPase activity

In order to down-regulate starch levels in transgenic plants, AGPase activity was decreased by the antisense RNA technique. The cDNA encoding the small subunit of the potato enzyme (Müuller-Röber at el., 1990) was inserted in its reverse orientation behind the cauliflower mosaic virus (CaMV) 35S RNA promoter and transformed into potato. The leaf AGPase levels of some of the transgenic plants were decreased to 5-30 per cent of the wild-type activity. A good correlation was observed between the amount of residual mRNA and protein expression and the degree of inhibition in AGPase.

8

Scopes and Achievements

Transgenic plants are widely grown all over the world. Most of these *first generation* products have specific agronomic traits designed to improve the efficiency of production. New varieties under test contain product quality or other higher value trails intended to provide specific benefits to end user. Examples of such *second generation* traits and attempts to predict range of household, medical, industrial products that might become available in coming years.

Most transgenic plants are generated for basic research purposes, such as the evaluation of tissue-specific expression of genes, processing and secretion signals, and enhancer elements. A growing effort is being made to use transgenic plants for broader objectives. A wide variety of foreign products can be obtained from transgenic plants, not only proteins, but also all kinds of modified compounds through the expression of the appropriate enzymes or even catalytic antibodies. The production of plant biomass is inexpensive, the plant cell machinery can provide specific processing unavailable from microbial systems, the isolated products are free of bacterial contaminants, the plant seeds can be used as low-cost storage, and ultimately plants use sunlight as their source of energy. These facts make us believe that the production of foreign compounds in plants is becoming a competitive alternative system to traditional methods.

OUR NEED

Crop improvement by engineering for specific traits can also be achieved through production of enzymes and metabolites encoded by foreign genes. Thus, the amino acid composition of seed proteins has been altered and, for example, transgenic canola, expressing a Brazil nut albumin, contains up to

33% more methionine in its seed protein (Altenbach et al., 1992). The alkaloid composition of some medicinal plants has been improved. The production of scopolamine, a medicinally important anticholinergic drug is highly increased in transgenic *Atropa belladonna* plants expressing hyoscyamine 6β-hydroxylase from *Hyoscyamus niger,* which catalyzes the conversion of the precursor hyoscyamine to scopolamine (Yun et al., 1992). The shelf life of fruits and vegetables can be extended. Tomato expressing a bacterial ACC deaminase gene reduces ethylene synthesis, and the mature fruits remain firm for at least 6 weeks longer than the nontransgenic controls (Klee et al., 1991). Flavor can also be enhanced, transgenic lettuce and tomato have been obtained that express monellin, a sweet protein isolated from some African berries that elicits a flavor approximately 100,000 times sweeter than sugar on a molar basis (Peñarrubia et al., 1992). The freezing properties of harvested produce can be improved and, in tomato plants expressing an antifreeze protein from the polar fish, winter flounder, ice recrystallization is inhibited (Hightower et al., 1991), making accumulation of the protein cold-specific (Kenward et al., 1993). The toxicity of some vegetables can be reduced. More than 70% of human cadmium intake is contributed by vegetable products. Shoots or transgenic tobacco expressing the mouse metallothionine I gene contain about 24% lower cadmium concentration than nontransgenic plants, because the pollutant metal is sequestered in the roots as cadmium-metallothioneine complexes (Yeargan et al., 1992.).

FUNGAL RESISTANCE

Plant diseases have been with us since agriculture began and intensification and continuous cultivation have resulted in increasing disease pressure. The epidemic spread of fungal diseases is controlled by various methods, like application of protective agrochemicals, breeding for resistance varieties and various crop husbandry techniques. Plant breeders have developed a large numbers of cultivars resistance to fungal diseases. The breeding programmes are time consuming, therefore, keep pace with the rapid evolution of the pathogen. The use of agrochemicals by the farmers is costly and eventually less efficient due to the evolution of the pathogen. It is therefore, necessary to develop disease control measures as alternative to chemicals. It is a priority of scientists worldwide. Newly developed technologies in plant breeding such as isolation of resistance genes and gene transformation methods can be used to develop transgenic plants. Furthermore, there is an extensive search for genes that encode enzymes involved in the synthesis of toxic to fungi and for genes encode proteins with direct inhibitory effect on the growth of fungi.

Fungal diseases cause world wide economic losses. Although plants have natural resistance mechanism but pathogens have some how also developed ways to evade these defence mechanism (Bowles 1999; Murray et al., 1993). Bowles (1990) grouped the defence genes into these classes.

1. Genes synthesizing compounds involved in cell wall modification.
2. Genes encoding defence related proteins exhibiting anti microbial activities or catalyse the synthesis of products that retard microbial activity.
3. Genes encoding pathogen related (PR) proteins whose appearance is correlated with defence responses.

In response to pathogenic fungi attack the plant synthesize defence related proteins which include amylase proteinase inhitiors, hydrolytic enzymes like chitinase and enzymes involved in synthesis of phytoalexin (Dickman et al., 1989; Murray et al., 1993). Phytoalexins (secondary metabolism product) are low molecular weight antimicrobial compounds which act as broad spectrum antibiotics play an important role in arresting the growth of fungal pathogens phytoalexins constitute a chemically heterogeneous group of molecules typical of plant secondary metabolites, which belong predominantly to the families of the phenylpropanoids isoprenoids and acetylenes (Bailey and Mansfield, 1982; Bailey, 1987; Bryngelsson et al., 1991). First phytoalexin to be identified and purified was pisatin from pea (Perrin et al., 1962). Direct proof that phytoalexins plays an important role in disease resistance was obtained from experiments in which the pisatin demethylase gene was transferred to *Cochioliobolus heterostrophus*, a pathogen of maize but not of pea unable to detoxify pisatin. It is predicted that plants able to make large amounts of foreign phytoalexins would be resistant to pathogens that could detoxify the new chemical structures. Resveratrol, a stilbene type phytoalexin, is produced in several plants, including *Vitis vinifera*, and its concentration has been shown to correlate with resistance to fungus *Botrylis cinerrea* (Stein and Blaich 1985). Stikene synthase, the key enzyme in the biosynthesis of Stilbenes, has been purified and characterized and genomic clones containing the gene encoding this enzyme have been isolated. Transgenic tobacco plants containing the stikene synthase (vit) genes were shown to produce resverarol and to be less susceptible to disease when infected with *B. cinerea.*

A common natural mechanism of plant to resists fungal attack is secretion of the enzyme chitinase which attacks the cell wall of fungus. This enzyme chitinase is very stable, resistant to heat and inhibits fungal growth in-vitro.

Seeds of various cereals contain proteins that are toxic to some pathogens. For example, barley contains a ribosome inhibiting protein (RIP) that prevents peptide elongation. This protein while not toxic to plants, inhibits the growth of number of fungi. Using a chimeric capable of expressing barley RIP in stem and roots of tobacco, Logemann et al (1992) developed transgenic plants resistant to *Rhizoctonia solani*.

Long lived fruits and vegetables

Fruits and vegetables are spoiled due to ethylene formation that triggers fruit ripening. Ripening can be delayed in transgenic plants. Tomatoes become soft due to production of polygalactouronase (PG) which causes break down of pectin in cell. Flavour and colour are not effected by this enzyme. DNA in anti sense orientation was inserted by Calgene scientists. Transgenic tomato exhibit decreased PG level upto 99% thereby increasing shelf life.

Race-Specific Resistance Genes

True patho-systems of the cultivar/race depend on the presence and function of resistance genes in the host and a virulence gene in the pathogen. The genetic inter-dependence of host and pathogen determines the outcome of the interaction. It was first postulated in Flor's *gene for gene hypothesis*. The race-specific resistance is characterised by rapid and strictly localised cell death in the host (hypersensitive response) limiting the spread of the pathogen to non-infected tissues and preventing its propagation. Various approaches are being pursued for isolation and molecular characterisation of resistance genes including map-based cloning and transposon tagging.

Transposon Tagging

It led to the isolation of a race-specific resistant gene from maize, the genetically defined HM1 locus' conferring resistance to race 1 of the fungus *Helminthosporium carbonum* (Johal and Briggs 1992). The derived amino acid sequence of the HM 1 genes shows similarity with the amino acid sequence of plant NADPH and NADH-dependent reductases and very likely the HM 1 gene product is responsible for detoxification of the HC-toxin. This pathogen-produced toxin mediates the specific pathogenicity of *H. carbonum race 1 on maize.*

After isolation of fungal resistance genes, they can be transferred to other cultivars/species to provide resistance to specific races of fungal pathogens. The resulting transgenic plant will remain susceptible to other races of the pathogens. Furthermore, the large scale growth of new bred fungus-resistant cultivars is likely to lead to evolution of new virulent races, and the cultivation of crops as monocultures will promote rapid epidemic outbreaks.

Two-Component Systems For Non-Specific Resistance

The durable non-specific resistance was obtained by engineering plants which have hypersensitive response is elicited by a wide range of pathogens

(P. de Wit, 1992). They further proposes a model for transfer of the avr 9 gene from fungus *Cladosporium fulvum* into a tomato lines containing resistance gene Cf 9. The model proposes that in transgenic plant containing this two-component systems, a hypersensitive response will establish in many host/pathogen interactions, if expression of the avr 9 gene is put under control of a promoter, there is rapid and local transcriptional activation in response to a broad spectrum of pathogens. Upon isolation of the Cf 9 gene, the avr 9-Cf 9 system could be introduced as a cassette also into plants that lack Cf 9. The strategy requires that the avr 9-encoded elicitor is targeted to correct cell within the transgenic plants to give a functional interaction with Cf 9 genes product. The success of this concept is largely dependent on the availability of suitable promoter to direct avr 9 expression.

Martini et al. (1993) developed a strategy for artificial generation of programmed cell death in transgenic potato, potentially limiting the spread of a wide range of fungal in the host tissue. A promoter fragment of potato PrP1-1 gene has been combined with barnase gene from *Bacillus amyloliquefaciens* encoding a highly cytoxic RNase. Rapid synthesis of this RNase in the vicinity of infection sites should initiate necrosis of host cell during early stage of compatible interaction and therefore, restrict the growth and propagation of biotrophic pathogenic fungi.

Compounds Toxic to Fungi

Microorganisms and plants produce low molecular weight antimicrobial substances. The plants produce phytoalexins and are often synthesised locally. They are accumulated after exposure to pathogens and/or stresses. In *Vitis vinifera* and *Picea sitchensis*, stilbenes are involved in protection against fungal challenge. Different biotic and abiotic signals have been shown to stimulate their accumulation. They activate genes for the key regulatory enzymes (phenylalanine-ammonia lyase and stilbene synthase) for their biosynthetic pathway.

Tobacco contains the precursors for the formation of stilbenes but lack the enzyme stilbene synthase. After transfer of a stilbene synthase gene from grapevine to tobacco transgenic plants expressed the foreign gene in response to treatment with fungal elicitor or infection with the broad range of fungal pathogen. The reduction of disease incidence on leaves directly correlated with amount of accumulated resveratrol with rapidity of stilbene synthase gene activation.

Anti-Fungal Protein

The characteristic features of fungal structure and cell metabolism represent potential targets for inhibitory agents. The constitutive or pathogen-inducible expression of genes encoding proteins with fungi toxic or fungistatic capacity in transgenic plants may enhance resistance to fungal damage in these plants.

Chitin and β-1, 3-glucans are structural polysaccharides of the cell wall of many fungi. The breakdown of these components by endochitinase and β-1, 3- endoglucanases may inhibit fungal growth. Four classes (1-4) of plant endochitinases have been described of which atleast three are present in tobacco (Melchers et al. 1993). Three major classes (1-3) of β-1,3-endoglucanases are also found in this plant species. Class 1 hydrolases are localised in plant vacuoles and exert a strong inhibitory effect on fungal growth in vitro. Inhibition is far more pronounced when mixtures of the two types of hydrolases are applied indicating that the proteins act synergestically. In contrast, class II hydrolases, which are similar to the class I hydrolases on the basis of their primary structure but are localised extracellularly, are not antifungal either alone or in combination with other proteins (Melchers et al. 1993).

Broglie and coworkers (1991) combined the bean endochitinase CH5B gene encoding a class 1 protein, with the promoter region of the CaMV 35 S gene conferring high constitutive expression in a wide variety of plant cell type. In the non- infected stage, transgenic homozygous tobacco plants harbouring this chimeric construct showed a 2 to 4 fold increase in the roots and 23 to 44 fold increase over control plants in chitinase activity in the leaves. For phytopathological evaluation, these plants were grown in the presence of *Rhizoctonia solani*. Seedling mortality or the loss of root fresh weight was clearly reduced in transgenic plants compared to control plants. When 35 S-chitinase plants were infected with the non-chitinous fungal pathogen *Pythium aphanidermatum*, no difference in survival was detected compared to control plants.

In contrast, Neuhaus and coworkers (1991) found that transgenic tobacco plants bearing a class I chitinase from tobacco under the control of the CaMV 35 S promoter did not reveal a substantial increase of resistance to the chitin containing fungus *Cercospora nicotiana*, although most transformants expressed high level of enzymatically active chitinase in the intracellular compartment of leaves. This could indicate that factors other than chitinase appear to be limiting in defence against this pathogen.

The group of J. Ryals at CEIBA-GEIGY, Research Triangle park (U S A) has reported on protection of transgenic tobacco plants against various fungal pathogens as a result of high- level expression of proteins associated with systemic acquired resistance (SAR) in this plant species. The majority of these SAR proteins belong to group of pathogenesis-related (PR) proteins. Their mRNA accumulates not only in pathogen- infected (primary) leaves but also in non-infected (secondary) leaves. The goal of this work is to achieve broad range protection against pathogen by over-producing combination of SAR proteins in transgenic crops.

Ribosome-Inactivating Proteins (RIPs)

Plant RIPs inhibit protein synthesis in target cells by N-glycosidic cleavage of 28 S rRNA (Stirpe et al. 1992). Two types of RIPs exist-single chain type 1 proteins and type 2 proteins consisting of two chains can bind to cell surfaces. The best known is ricin and is most potent natural toxin. RIPs do not interfere with the functioning of self ribosomes but exert inhibitory activity towards fungal ribosomes. In *in vitro* assays a barley RIP has a lower antifungal activity than chitinase or B-1,3-glucanases from barley. However, a strong synergy is observed when barley RIP is mixed with either of these two hydrolases (Leah et al. 1991). Logemann et al. (1992) improved resistance to *R. solani* in transgenic tobacco. A barley type 1 RIP gene was put under control of the potato Wun1 promoter which mediates wound and pathogen- inducible transcription in the epidermis of leaves, stems and root from tobacco. R1 progeny showed an increased resistance to *R. solani*.

In the mean time , a large number of proteins with inhibitory effect on fungal growth *in vitro* have been identified from the plants. Among these are osmotin which are produced in response to salt stress and cysteine-rich proteins of various origin. The function of these proteins in transgenic plants remains to be tested.

The transgenic plants show enhanced resistance against specific fungi, although such crops have not reached the market yet. New concepts have been developed to obtain broad range pathogen resistance, and these concepts will be worked out further in the next few years. Undoubtedly, the molecular basis of both pathogenicity and resistance will lead to the development of new strategies in the future.

ENGINEERED RESISTANCE AGAINST VIRUS DISEASES

Plant viruses cause varied diseases and significant crop losses due to the lack of proper *viricides.* Virus diseases are usually, controlled by using certified virus free planting material, by eradicating infected plants and by spraying chemicals against virus vectors. These measures are not very successful.

The genetic engineering of virus resistance into plants provide a direct and more efficient control of virus diseases. The engineering strategies are based on the introduction of a virus-derived or a virus-targeted sequence into the genome of a potential host plant. These sequences interfere with specific stages in the viral infection cycle.

The genomes of viruses consist of single or double-stranded RNA or DNA. The vast majority of pathogenic plant viruses (about 80 %) have genomes of plus strand (messenger sense) single stranded RNA but also viruses with minus sense or ambisense genomic RNA (TSWV), single-stranded DNA (geminiviruses) or double-stranded DNA (caulimo and badnaviruses) cause significant losses to crop plants. All viruses encode proteins involved in the replication of their nucleic acid. RNA viruses encode a putative RNA-dependent RNA polymerase, while double-stranded DNA viruses, which replicate via RNA intermediates code for reverse transcriptase. Single-stranded DNA viruses contain genes for enzymes necessary for replication via complementary DNA. Besides the replication enzymes, genome of most viruses also encodes a coat protein, which protects the viral nucleic acid during the transfer of virus particles from one plant to another. Additionally, many viruses synthesize proteins that facilitate their movement from cell to cell in the host plant and there are also several viral gene products with unknown function. To enable monocistronic translation, some viruses have divided their genome into segments, each representing a mRNA for one protein, others synthesize sub-genomic RNAs which are mRNAs for proteins encoded downstream of the 5- terminal cistron. Finally some viruses translate their genôme into polyprotein, which is processed into functional proteins by virus-encoded proteases.

To engineer resistance into host plants the following strategies are currently under investigation :

Coat Protein-Mediated Protection (CPMP)

Coat protein-mediated protection concept is based on the observation of cross protection and was first used as a test of relatedness between viruses, but soon thereafter became a tool to control virus diseases in agriculture. A plant infected by a mild strain of a virus resists infection by other related strains, a phenomenon referred to as cross protection (Mckinney, 1939). The tobacco mosaic virus (TMV) coat protein (CP) is specifically involved in cross protection (Sherwood and Fulton, 1982).

Coat protein-mediated resistance to viruses is done through plant genetic engineering. Several major crop plants have been engineered to resist important viral pathogens and, in some cases, the resistant plants have been in field trials for several years. Beachy (1986) introduced the TMV coat protein gene into the genome of tobacco plants. Transgenic plants expressing high levels of coat protein upon infection with TMV exhibited a delay in disease development. Subsequently, CP genes isolated from different single-stranded (plus) sense RNA plant viruses have been cloned into plant expression vectors and transferred into various crop plants (tobacco, tomato and potato) using the *Agrobacterium tumefaciens*-mediated plant transformation system.

The CPMP approach has been extended to cereals (Toriyama et al. 1992). CPMP is achieved by transforming plants with a coat protein genes, under the control of CaMV 35 S promoter which mediated strong constitutive transcription in many cell types. The degree of resistance is correlated with amount of intact coat protein expressed in transgenic plants.

The introduction of non-functional transgenes is able to interfere with virus multiplication. Plant containing anti-sense coat protein RNA of CMV or PVX were protected against viral infection. High levels of resistance were obtained by introducing a coat protein cistron of PVX lacking a translational start signal (Goldbach et al. 1992). An anti-viral effect was also established by introducing untranslatable coat protein genes of TEV into tobacco plants (Dougherty et al. 1992). It has been suggested that resistance resulted from hybridization of the coat protein RNA to the minus sense replication intermediates of the viral RNA, thus inhibiting the production of infectious virus particles. The formation of RNA: RNA hybrids was also postulated as a mechanism inhibiting multiplication of TSWV, a virus with an ambisense RNA genome. Inhibition was mediated by the RNA of the nuleocapsid protein which encoded in viral complementary sense on a genome RNA fragment (Goldbach et al. 1992).

The results from these and several field trials demonstrate that CPMP might be very effective in preventing losses due to virus infection, but a large number of clones may need to be screened to identify the most resistant line. A better understanding of the mechanism (s) of CPMP should help to extend resistance beyond its current limitations.

Defective Interfering RNA/ DNA Protection

Defective interfering (DI) particles are deletion mutants of genomic viral sequences, which depend on their parent virus for replication. DIs RNAs have been described in animal virus systems, but only rarely in plant virus systems such as TCV, TBSV, CyRSV and TSWV. The occurrence of a DI viral DNA was reported for ACMV, a single-stranded DNA virus which produces sub-genomic DNA particles.

Like satellite RNAs, DI particles can attenuate the disease symptoms of their helper virus by interfering with its replications. Thus, DI RNAs or DNAs may also be used as tools for genetic engineering of plants. There have been attempts to construct artificial DI molecules of TYMV, TMV and BMV. For example deletion mutants of BMV RNA 2 were shown to reduce the replication of BMV RNA 1 and 2 in barley protoplasts (Hall et al. 1991).

Satellite RNA-mediated Disease Attenuation

Satellite are small extragenomic RNA species that depend on a helper virus for replication. Some satellite RNAs are known to attenuate the symptoms of their helper virus. Transgenic tobacco plants expressing the satellites of CMV or TobRSV were shown to be protected against severe effects of their respective helper virus. The expression of CMV satellite RNA also protected against the symptoms of infection with TMV, but not against a range of other viruses (Baulcombe et al. 1987).

Anti-sense RNA/ Ribozyme-Mediated Protection

Antisense constructs have generated protection against RNA and DNA viruses. Antisense coat protein genes mediate high level of resistance in PLRV. Huntley and Hall (1993) reported an antisense transcript of BMV RNA 3 intercistronic region interferes with viral RNA replication.

Ribozymes are RNA molecules catalyzing endonucleolytic cleavage reactions. Some viroids, virusoids and satellite RNAs perform self-cleavage reactions. Mostly, the cleavage site consist of a consensus structure, called the "hammerhead" motif. The nucleotide region directing the catalysis of cleavage reaction could be separated from the region where the cleavage occurs and the recognition of the target could be modified by changing the nucleotide sequence of the regions flanking the cleavage site. Steinecke et al. (1992) designed a ribozyme targeted against the mRNA of the npt gene, which was shown to reduce mpt gene expression when transiently expressed in tobacco protoplasts. A cleavage product of the NPT mRNA was detected.

Replicase and Protease-mediated Protection

Golembowski et al. (1990) discovered that transgenic tobacco expressing part of replicase gene of TMV strain was highly resistant to this and closely related strains. In contrast, transgenic tobacco expressing functional AIMV replicase protein was not virus resistant and in barley protoplasts, intact BMV replicase did also not reduce virus replication (Morei et al. 1992). Longstaff et al. (1993) changed the conserved NTP-binding or polymerase motifs of the PVX open reading frame 1 RNA (encoding the putative viral replicase) rendering the viral genome non-infectious. The mutant genes were transformed into tobacco and were then shown to confer extreme resistance against PVX infection.

The recent report describing virus resistance in plant expressing protease genes of polyviruses suggest that this could be an additional new type of pathogen-derived resistance based on interference with the processing of the viral polyproteins (Vardi et al. 1993). This possibility is not yet proved.

Movement Protein-mediated Protection

The movement of viruses between cells is an attractive target for transgenic resistance mechanisms. It has been expected that resistance based on cell to cell movement might have broad spectrum effectiveness. There are reports that virus resistance in transgenic tobacco is mediated by a defective movement protein of TMV (Lepidot et al. 1993). It is shown that the dysfunctional protein responsible for resistance does not modify plasmodesmata. The movement of protein- mediated resistance was effective, at least partially, against virus of the TMV group that were only distantly related to TMV. Transgenic expression in tobacco of the BMV movement protein also conferred partial resistance to TMV. It might be considered that the movement-protein of BMV dysfunctional in tobacco.

Resistance through Transgenic Expression of Antiviral Proteins

There have been reports of virus resistance resulting from transgenic expression of genes encoding a single chain antibody against a virus encoded protein. By constructing transgenic plants expressing single chain antibody gene Tavladoraki et al. (1993) have shown specific protection against artichoke mottle crinkle virus. This report is the first example of resistance to pathogen attack through expression of transgenic antibodies directed against a structurally important epitope of a viral coat protein. The resistance conferred by transgenic expression of rat 2-5 oligo A synthetase in potato plant is effective against two unrelated viruses PVX and TMV (Truve et al. 1993). Precisely how this resistance might operate is a mystery.

The pokeweed anti viral protein is a ribosome- inactivating protein similar to ricin and related proteins found in many plants. These proteins inactivate ribosomes by de-adenylation of ribosomal RNA and are thought to contribute to natural defense by serving as cellular suicide proteins. Normally, the proteins are located in the cell wall. They are released into the cell when the plasma membrane is broken by physical damage (when plants are penetrated by invertebrate virus vectors or when the cells are mechanically damaged in a manner that allows virus entry). A recent report (lodge et al. 1993) indicates that tobacco plants transformed to a virus-resistant phenotype by transgenic expression of the pokeweed anti viral protein.

Following are the approaches to broad spectrum resistance:

1. The first is based on the increasing molecular understanding of the function of virus encoded proteins mainly implicated in replication and cell to cell movement of virus.

2. A second approach relies on the transgenic expression of antivirus protein of non-plant origin including antibodies.
3. The third approach is based on genetic manipulation of natural viral defenses in plants.

Over the recent years, progress has been made towards identification of viral components that trigger these resistance mechanism. In the near future there is good prospect that the resistance genes that control these defense mechanisms will be cloned and characterised. With the combined information about these host and viral components, it should be possible to design methods of manipulating the broad spectrum component of the natural defense mechanisms.

Table 8.1: Examples for protection against virus infection mediated by coat protein sequences

Source of Coat Protein gene	Transformed plant species	Transgenic expression protects against
TMV	Tobacco	TMV
TMV	Tomato	TMV, ToMV
TMV	Tobacco	PVX, CMV, AIMV, SHMV
AIMV	Tobacco	AIMV
AIMV	Tobacco	PVX, CVM
AIMV	Tomato	AIMV
AIMV	Alfalfa	AIMV
TRV	Tobacco	TRV, PEBV
CMV	Tobacco	CMV
PVX	Tobacco	PVX
PVX	Potato	PVX
PVX+PVY	Potato	PVX+PVY
PVS	Potato	PVS
PLRV	Potato	PLRV
TSWV	Tobacco	TSWV
TEV	Tobacco	TEV
RSV	Rice	RSV
GCMV	Tobacco	GCMV

For abbreviations, see in end of the book.

HERBICIDE TOLERANCE GENES

Expression of foreign gene conferring herbicides in transgenic plants lead to high level of herbicide resistance. Herbicide tolerance can be obtained by following methods:

Herbicide target modification

Herbicide targets are proteins and their action is two folds: it either inhibits photosynthesis or amino acid biosynthesis.

Triazines are common herbicides which inhibit photosynthesis.

1. Triazines block electron transport in PS-II by binding to Qb protein present in thylakoid membrane encoded by *psbA* gene of chloroplast DNA (Zurawski et al., 1982).
2. Triazine resistant mutants having one altered amino acid have been identified from naturally occurring plants or microbial species (Ohad et al., 1990).
3. Triazine resistance can be obtained by transferring the resistant chloroplast genome by sexual hydridization or by developing a chloroplast transformation system.
4. Tobacco cells were transformed by *Agrobacterium* vector having a pabA gene coding a triazine insensitive protein, fused to transist peptide of nuclear encoded chloroplast protein (Cheung et al., 1998).

Triazines which inhibit amino acid synthesis:

1. Triazines affect amino acids biosynthesis, like sulphonyl ureas, imidazolinones.
2. The target enzyme is acetolactate or ALS (La Rossa et al., 1984; Ray, 1984; Shaver et al., 1985).
3. Mutant forms of ALS resistant to herbicides having one amino acid sequence have been identified and isolated.
4. Maize plant tolerant to imidazolines correlated with the presence of altered ALS have been produced (Shaver et al., 1985).

Over production of target enzyme

Glyphosate (Round up™) interferes with amino acid biosynthesis by inhibiting the enzyme EPSP (5-enol pyruvyl shikimate 3-phosphate synthase). High level herbicide tolerance has been obtained in plants by over production of EPSP in chloroplast (Shah et al., Shah et al., 1988). *aro A* bacterial gene was transferred from *Salmonella typhimurium* into tobacco (Comai et al., 1985; Fillatti et al., 1988). Transformed plants showed resistance to glyphosate. Tolerance is dependant on *aro A* encoded bacterial EPSP.

Detoxifying enzymes

Herbicides detoxifying enzymes inactivate herbicides before they are able to inhibit the target enzyme. The most successful example is the detoxification of phosphinothricine phosphate (Murakami et al., 1986).

Table 8.2: Herbicide Resistant Transgenic Plants

Species Modified	Transgene Source	Transgene Product
Target modifications:	*Arabidopsis thaliana*	Acetolactate synthase
Beta vulgaris	*A. thaliana*	Acetolactate synthase
Nicotiana tobaccum		
Enzyme over production:	Plant and microbial genes	Analogue of EPSP synthase
Glycine max		
Enzyme detoxification: *Brassica napus*	*Streptomyces hygroscopicus*	Phosphinothricin

Insect tolerance

Pesticides are often highly toxic to non-target organism. Transgenic plants can be developed that are able to protect themselves against insects by expressing insecticidal protein or a proteinase inhibitor gene. *Bacillus thuringiensis* produces insecticidal protein crystals during sporulation that are toxic to lepidopteran insects. Commercial insecticide of this bacteria is effective against 20 lepidopterans. Fischoff et al (1987) constructed chimeric genes containing 35-S promoter of CaMV and protein coding sequence of bacterium. Plant can be made insect resistant by introducing genes for certain protease inhibitors. Proteins as inhibitors conferring endogenous resistance to insect attack are common in higher plants. In addition, they have anti-metabolic activity in wide range of insects.

Coat protein genes for virus protection

Control of plant virus depends on methods that prevent or restrict its infection. Most successful method of producing virus resistant plant is to insert and allow expression of a viral coat protein genes. The principle is based on the observation that infection of a plant with one viral strain protects against super infection by another related strain. This phenomenon is called "Cross-protection". It is assumed that the mild virus which infected the cell first, produces inhibit the uncoating of RNA of second virus thereby delaying or preventing the RNA expression and replication (Hilder et al., 1990). Powel – Abel et al.(1986) were first to describe the expression of TMV coat protein when infected with TMV in transgenic tobacco. A C-DNA coding for TMA coat protein under the control of CaMV 35-S promoter was transferred to tobacco. Progeny showed delayed symptoms or no symptoms.

Using a non structural gene from TMV strain U1, Golemboski et al (1990) were able to produce highly virus resistant transgenic tobacco plants, the chimeric gene used contained a CaMV 35-S promoter. Tansgenic plant showed complete resistance even when inoculated with high concentration of virus.

Virus resistance can be obtained via satellite RNA expression. Satellite RNA's are small RNA molecules encapsulated by plant viruses and packaged together with viral genome. Certain plant RNA viruses harbour these molecules. Such molecules are unable to replicate independently but require assistance from the virus. The transgenic plants which expressed the satellite RNA when infected with corresponding virus exhibit significant delaying symptom development compared to the plants devoid of gene coding RNA.

Light regulated genes

Many light regulated genes show tissue specific expression. Transgenic plants aid in analyzing such genes. Cis elements for tissue specific expression of light regulated genes have been investigated. Transgenic tobacco plants containing the chalcone synthase(CHS) promoter from *Antirrhinum majus* and NPT-II structural sequence was obtained (Kaulen et al., 1986). The transcription of a chimeric gene with promoter was shown to be light inducible in tobacco.

Antisense RNA

This technology involves the reduction in gene expression by using RNA. The proposed mechanism is the formation of ds RNA that is readily degraded than free m-RNA. Rothstein et al(1987) demonstrated inhibition of nopaline synthase (NOS) gene in tobacco. This method can be used to reduce ripening in fruits.

TRANSPOSABLE ELEMENTS

In 1950's, Barbara Mc Clintock proved that genetic elements exist, such as activator (Ac) in maize, which can transpose and generate unstable mutable alleles. Transposon tagging is used for isolating genes which encode unknown products. Number of genes have been cloned from maize and *Antirrhinum* by molecular tagging using transposable elements.

SEED STORAGE PROTEINS

In legumes and most of the dicots, the major storage proteins are salt soluble globulins. In monocots, prolamines (alcohol soluble) and glutelins (acid-base soluble) are dominant seed proteins. Tissue specific expression of these genes have made it possible to clone and transfer these genes into various plants. Transgenic tobacco plants have been obtained by introducing 917 Kbp soyabean DNA fragment containing lectin gene and at least four non-seed protein genes into tobacco genome (Okamuro et al., 1986).

BIOACTIVE PEPTIDES

Biologically active peptides are chemically synthesized or obtained by extraction from biological materials or by expression as fusion proteins in prokaryotic or eukaryiotic cells (Josephson and Bishop, 1988). Leu-enkephalin, a pentapeptide (YGGFL) with opiate activity has been produced in *Arabidopsis* and oilseed rape as part of chimeric plant seed storage proteins.

The 2S albumins, or napin-like proteins, are formed by two subunits linked by disulfide bridges (Lonnerdahl and Janson, 1972) and are synthesized as prepropolypeptides, being co- and post translationally processed (Chrispeels et al., 1982) and stored in the protein bodies. The 2S albumins from different species are structurally very similar, but contain a region between the sixth and seventh cysteine residues that is highly variable (Ampe et al., 1986).

ANTIBACTERIAL PEPTIDES

Lytic peptides are small proteins with antimicrobial activity that appear to be major components of the antimicrobial defense system of a number of organisms, including insects, molluscs amphibians and at least one mammal (Lee et al., 1989). These peptides are generally sequences of 23 to 39 amino-acids which have the potential for farming amphipathic $\propto$-helices. Cecropins (lytic peptides) are a family of small, basic polypeotides present in the haemolymph of insects. The structural features of cecropins are necessary for their antimicrobial activity, a process involving channel formation and subsequent membrane disruption (Christensen et al. 1988). To test the use of cecropins in enhancing resistance against bacterial pathogens in transgenic plants, Jaynes (1993) produced transgenic tobacco plants that harboured c!imeric genes, with the coding sequence of a synthetic homologue of vcecropin B placed under the control of the potato proteinase inhibitor II promoter. Transgenic plants had less wilted leaves and significantly lower mortality than control plants.

HUMAN PROTEINS

Human serum albumin (HSA) has been expressed in transgenic potato (Sijmons et al., 1990). It is protein of broad clinical use and, consequently, of a high commercial value. In humans, it is synthesized as preproalbumin, released from the endoplasmic reticulum after removal of the prepeptide (Judah et al., 1973) and proteolytically processed by a serine protease that cleaves the six residues of the propeptide (Brennan et al., 1984). Similarly to HSA, secreted plant proteins contain NH_2-terminal signal peptides that are removed from the preprotein by a specific protease. The NH_2-terminal amino acid sequence analysis of purified HSA produced in transgenic potato plants showed that the processing of the protein depended on the type of signal sequence. The two signal peptides are capable of targeting HSA extracellularly. The HSA accumulated up to 0.02% of the total protein. The amount could probably be increased in tubers by using a tuber-specific promoter (Wetzler et al., 1989). HSA added to round untransformed potatoes did not show any degradation during the industrial process of starch extraction (Sijmons et al., 1990), as it was the protein obtained in the waste liquid fraction of the process. This observation suggests that transgenic starch tubers could well be used as an inexpensive source for protein production, provided that the levels of HSA could be increased. Human interferons (INFs) are small defense proteins related to resistance to viral infection. They are high-value proteins and have been cloned and produced as recombinant proteins in bacterial and cultured animal cell systems. Interferon-like factors have been detected in virus-infected plants (Sela, I.), and some have been isolated by their affinity for antibodies against human interferon beta, although they do not show structural homology to the interferons (Edelbaum et al., 1991). Human interferon β has been expressed in tobacco transgenic plants (Edelbaum et al., 1991). It was detected by immunoblot analysis and extracts from positive plants inhibited the cytopathic effect of vesicular stomatitis virus in human wish cells. Transgenic plants producing interferon-β could be used as an inexpensive system to produce the protein.

ENZYMES

They are used, soluble or immobilized in a wide variety of industries. They are isolated from the organisms that produce them or they are obtained as recombinant proteins. Transgenic plants may constitute an appropriate system to produce industrial enzymes. Hoekema's group have expressed *Bacillus licheniformis* α-amylase in transgenic tobacco (Pen et al., 1993). It is used in the starch-processing industry, and also in others, such as the brewing, baking, wine and juice, and detergent industries. *Bacillus licheniforms* α-amylase is the most commonly used in starch liquefaction because of its head stability and wide pH range of activity. The molecular mass of the enzyme produced in tobacco was 64 kDa; this value is higher than that of the *B. licheniformis* enzymes, which is only 55.2 kDa. This difference could be attributed to glycosylation in plants. Despite differences in glycosylation, the enzyme exhibited biological activity and had similar thermostability. Hoekema's laboratory has expressed *Aspergillus niger* phytase in transgenic tobacco (Pen et al., 1993). Phytase catalyzes the conversion of phytate (*myo*-inositolhexaphosphate) into inorganic phosphate and *myo*-inositol. Phytate is the main storage form of phosphorus in many plant seeds used as animal feed. A phytase from *A. niger* has been developed that functions well in the acidic conditions found in the gastrointestinal tract of these animals. Transgenic phytase seeds were tested *in vivo* by milling and subsequent addition to the basal diet of broilers.

ANTIBODIES

Of the variety of bioactive compounds expressed in transgenic plants, antibodies may offer the widest range of possible applications. Antibodies produced in plants possess all of the characteristics and are virtually indistinguishable from antibodies produced by hybridoma cells. Recently, it has been found that antibodies that possess catalytic capabilities can be isolated from mouse hybridoma et al., 1990). Genetically stable seed stocks of antibody-producing plants can be isolated and stored indefinitely at low cost; the seed stock can be converted into a harvest of any quantity of antibody within one growing season. Although tobacco has been used as the principal research tool to initiate the study of antibodies in plants, there may be more appropriate plants for production. A variety of common crop plants can be used as the production host. Acreages of perennial forage crops could be generated by clonal propagation or from seed and harvested numerous times in a growing season. The choice of species may depend on the quantity and nature of contaminants encountered during purification. Some candidates are alfalfa, soybean, tomato, and potato. Because large-scale production of antibodies is not yet commonplace, appropriate techniques for purification of hundreds or thousands of grams have yet to be perfected.

VACCINES

Plants represent a less expensive production system here, being most appropriate for oral vaccines that could be expressed in edible plant tissues. Hepatitis B surface antigen (HbsAg), which is used as vaccine against hepatitis B, has been successfully expressed in transgenic tobacco plants (Mason et al., 1992). Tobacco was transformed with constructs containing the sequence encoding for the HbsAg preceded by an enhanced CaMV 35S promoter and the 5'-untranslated leader of tobacco etch virus (Carrington and Freed, 1990). The presence in transgenic leaf extracts of material that reacts specifically with monoclonal antibody against HbsAg was tested, and levels ranging up to 66 ng/mg of soluble protein were found. However, if these levels could be increased, the inexpensive plant-produced protein could be used in vaccination against hepatitis B.

INDUSTRIAL PRODUCTS

Plants can also be engineered to produce foreign metabolites by introducing genes coding for the appropriate enzymes that convert compounds that are normally present in the plant into products of higher commercial value. Cyclic oligosaccharides of six (α), seven (β) or eight (γ) α-1,4-linked glucopyranose molecules are used in pharmaceutical industry as delivery system and in the food industry for flavour and odour enhancement and for removal of undesired compounds such as caffeine. Cyclodextrins are prepared *in vitro* by the action of the bacterial enzyme cylodextrin glycosyltransferase (CGT) on prehydrolyzed starch. The gene encoding cyclodextrin glysocyltransferase from *Klebsiella pneumoniae* has been cloned and has been expressed in transgenic potato. A tuber-specific promoter from a class I patatin gene was used to restrict expression of the enzyme to the tuber, which storage of large amounts of starch, the CGTs substrate. Northern blot analysis of the transformants showed transcription of the chimeric gene, but no CGT was detected in tubers by immunoblot analysis. A potato tuber contains approximately 14% starch, of which 0.001-0.01% was converted to cyclodextrins. For the commercial production of cyclodextrins, this value should be increased to 1-10%. This might be possible by using stronger promoters and more efficient targeting signals.

Plastics are probably the most used materials in industry nowadays. Their use created environmental problems, because most of them need extremely long period to biodegrade. Therefore, growing efforts are being made to obtain inexpensive biodegradable plastics.

METAL RESISTANCE IN PLANTS

Metals like aluminium when present as ions poison plant roots by making cells rigid and unable to lengthen. *Arabidopsis* plant after random mutations (by chemicals) were screened for A1 tolerance. One of the mutants coped with A1 by secreting organic acids (citric acid and malic acid) which bound the ions outside cell forming respective salts and thereby inhibiting their entrance into root tip cell. The gene responsible is mapped to be present on first chromosome of *Arabidopsis*. The second type of mutant responded by increased flux of hydrogen ions into root tip, alkalinizing the medium outside root. Increase in external pH (0.15 unit) leads to formation of aluminium hydroxides and precipitates that failed to enter cell. Other metals like Cd and Cu produce free oxygen radicals. Plants cope with these metals by binding them in complex with peptides called phytochelatins. Experiments show that genes for a metal resistance enzyme called phytochelatin synthase from wheat, yeast and *Arabidopsis* enable yeast to survive a dose of Cd that kills control cells.

TRANSGENIC PLANTS WITH HIGHER STRESS TOLERANCE

To increase productivity it is necessary to increase tolerance against abiotic stresses. Environmental stress after gene expression (Sachs et al., 1986) and permits isolation of stress related genes. Plants can be genetically engineered to improve stress tolerance. Many organism have gene that enable them to survive in extreme environments and which can be introduced in higher plants. Stress relieving genes might encode enzymes involved in particular metabolic pathway. regulatory or protective proteins. Thus stress responses help organism in adaptation by changing metabolic flow.

Reduction in water availability causes reduction and failing leads to complete inhibition of photosynthesis. In condition when photosynthesis is impaired and chloroplast are exposed to light there is photoreduction of oxygen forming products like peroxides, superoxides that damage membrane system. Anti oxygen enzyme systems scavange reactive oxygen intermediates. The three common types of stress that plants face are:

Water Stress

Many plants depend on their ability to function under water scarcity. Abscissic acid (ABA) concentration effects yield under water stress condition (Milborrow, 1981; Sachs et al., 1986). This forms an important criterion for drought resistance.

Temperature Stress

It includes heat tolerance, chilling stress and freezing stress. The reproductive stage is most sensitive to heat leading to floral abscission, pollen sterility and poor fruit set. Chilling is the most severe environmental stress that reduces germination and growth rate. Genetic resources for chilling tolerance have been found in maize (Mock et al., 1972). In crop plants considerable genetic variability occurs in freezing tolerance below zero degree centigrade. Frost hardiness is considered a quantatively inhierited trait controlled by a complex interaction of several genes (Thomashaw, 1990).

Salt Stress

Salinity causes water stress from the osmotic effects, mineral toxicity of salt and interruption of mineral nutrition of plants. Efforts to develop salt resistant varieties have been unsuccessful due to this multi partite nature of stress.

Strategy for Stress Tolerant Transgenic Plants

Basic steps involved in formation of stress tolerant plants by genetic engineering can be explained by following flower chart:

Search for organism which live under extreme environmental stress condition

↓

Study the organism under stressed and unstressed condition

↓

Isolate and clone gene/s that confer stress tolerance

Characterize the gene product in a prokaryote model system

Introduce the gene into plant system and study under stress

↓

Introduce gene in crop and perform field trials

Fig. 8.1: Basic steps involved in formation of stress tolerant plants by genetic engineering.

Properties/Characters Required for Improved Stress Tolerance

Stress tolerance can be improved by introduction of genes encoding enzymes that catalyze conversion of naturally occurring substrate into product with osmoprotective properties. Thus, properties that should be emphasized are:

Production of osmoprotective compounds

The ability of plant for accumulating low molecular mass osmoprotectants and osmolyts (amino acids, sugar alcohols) is a strategy to determine its resistance to environment stress. For example, mannitol (sugar alcohol) is produced in *Arabdiopsis* and tobacco by bacterial mltd gene encoding mannitol-1-dehydrogenase (Thompson et al., 1995; Tarczyski et al., 1993). Seeds of transgenic *Arabdiopsis* showed greater germination in high salt concentration.

Improved membrane flexibility

Chilling sensitivity depends upon the unsaturation of fatty acids in plastid membrane (Murata et al., 1982). Presence of centrally positioned cis-double bonds in the membrane lipids lowers phase transition period to zero degree centigrade. By introducing enzymes catalyzing formation of cis double bonds in saturated fatty acids, chilling resistant phenotypes can be obtained.

Stress induced proteins

Transcription of genes encoding late embryogenesis abundant (LEA) proteins is activated during abiotic stress. The protective effect of these proteins was improved by introducing HVA-I gene, encoding three LEA proteins from barley to rice (Xu, D. et al. 1996). Plants showed improved osmotic stress tolerance as indicated by delayed development of damage symptoms and improved recovery.

Scavenging reactive intermediates

Abiotic stress causes formation of reactive oxygen intermediates that cause oxidative stress by damaging membranes and macromolecules (Allen, 1995). Freezing and salinity resistant plants have well developed antioxidant defenses that scavange oxygen radicals.

Table 8.3: Showing foreign genes expressed in transgenic plants

Gene	Origin	Host	Stress
Bet A	*E. coli*	Tobacco	Salinity
Bet A	*E. coli*	Potato	Freezing
Mltd	*E. coli*	Arabidopsis	Salinity
Fad 7	*Arabidopsis*	Tobacco	Chilling
HVA-I	*Barley*	Rice	Salinity and drought
Mn-SOD	*N. plumbaginif olia*	Alfalfa	Drought and freezing

One approach for engineering stress tolerance may be to introduce genes from different stress responses to a single plant. It is important to target the location, control of time and level of expression and ensure precursor availability for each enzyme in order to avoid negative effects.

ALTERATION OF OIL COMPOSITION AND CONTENT WITH REFERENCE TO TRANSGENIC RAPESEED

Glycerol molecule with a fatty acyl group esterified to each of the three carbon (triacylglycerol) form the basic structure of seed oils. Fatty acyl groups are also essential components of structural lipids in plants. Some seed oils have unusual fatty acids (Hilditch and Williams, 1964); types found exclusively in Triacyl glycerols (TAG) of seed storage tissue, embryo or endosperm. An oil seed can be engineered to increase oil content if its genetics for oil contents is understood. Thus, it can be said that:

1. Fatty acid composition and seed oil content is genetically determined, genes determining useful traits can be identified and cloned.
2. Seed oil composition and content is due to expression of genes that are tissue specific.
3. Seed storage lipids in terms of fatty acids composition can be genetically engineered to some extent.

Seed oils can be genetically modified by modulating expression of genes of endogenous lipids synthesis pathway or by adding foreign genes. For changes in oil contents, pathway can be altered at following steps;

1. The initiating step which controls the accumulation of lipids.
2. FAS (fatty acid synthesis) cycle where changes may effect chain length and saturation.
3. Termination step where fatty acids form glycerolipids.

Over expression of gene can be obtained by an increase influx through a step; on the other hand a decrease in flux through antisense expression of a gene might be possible.

Gene transfer experiments help in identifying enzyme which controls the rate of lipid biosynthesis i.e the effect on oil content of over expression of genes encoding specific enzymatic activities may be examined *in vitro* in transgenic plants. Enzymes involved in the FAS cycle are parts of a multi-enzyme complex with a defined stoichiometry. This may cause hinderance in an attempt for their over expression. Over expression of one could result in proportional changes in other proteins. Reconstruction expression *in vitro* in transgenic plants can be useful in such case. Thus, antisense control of FAS enzymes level appears more feasible than increasing oil content by genetic manipulation. This can be explained by antisense expression of gene for stearoyl ACP desaturase. Reduced activity of enzymes in developing seeds leads to decrease in total unsaturates. Unsaturated fatty acids in most oil seeds including rapeseed are derived from conversion of stearate to oleate. This change is desirable as oil obtained has a composition similar to that of cocoa butter.

Table 8.4: Possible modification of rapeseed oil

Crop	Enzyme	Product	Application
Castor bean	Hydroxylase	Ricinoleic acid	Lubricant,laxative
Umbellularia californica	Thioesterase	Lauric acid	Detergents.
Nasturtium	Acyl transferase	Trierucin TAG	Plasticizers
Cacao	Acyl transferase	Cocoa fat equivalent	Specialty food.

Assuming that rapeseed or other crops can be genetically engineered to generate novel ratios of specific fatty acids, the TAG synthesizing machinery of host plant must be able to utilize such fatty acyl substrates.

FACTORS AFFECTING GENE EXPRESSION

Overexpression of genes involved in the control of rate of biosynthesis can increase oil content. In some cases, introduction of additional copies of endogenous gene may interfere with gene expression, thereby rather than enhancing the gene product. This phenomenon is called 'co-suppression'. Lowering gene expression by the use of antisense RNA has been successful in reduction of polygalactouronase levels in tomatoes (Sheehy et al., 1988; smith et al., 1988).

Regulation of gene expression

Lipids are essential components of plant membrane and serve other structural roles in all tissues of plants. Species that accumulate novel fatty acids in seed oil largely limit their presence to storage lipids. Thus, seed specific expression is desirable such that the structural lipid metabolism is not effected. Specificity even within seed (endosperm, cotyledons) may be required.

During development seed enters the maturation phase showing cell elongation and accumulation of storage reserves. It is a stage of maximum oil synthesis. Genes intended to alter fatty acid synthesis must be expressed during maturation phase when oil accumulation rates are maximum.

Regulatory sequences for the genes expressed in seeds during various stages could be used to express foreign genes in seeds. Genes for major storage proteins have been isolated and regulatory sequences from many of the genes have been tested in transgenic plants. Genes for major storage proteins have been isolated and regulatory sequences from many of the genes have been tested in transgenic plants. Genes isolated from "oil synthesis" which might be expected to contain regulatory sequences are those encoding ACP (acyl carrier protein) from *Brassica napus* (de Silva et al., 1990) and *B. campestris* (Scherer et al., 1992).

Genes expressed during seed development have been characterized including seed storage protein (napin and cruciferin). Regulatory elements from *B. campestris* genes have been characterized and expression cassettes have been developed that will be useful in expression of genes for modification of oil synthesis. Napin gene regulatory sequence are preferred as it is expressed earlier and corresponds to the period of oil accumulation in *Brassica.* Napin is encoded by a multigene family in *B. napus* and *B. campestris.* The upstream regulatory sequence of one of these genes, *gNa* is used to express a tagged napin gene in transgenic *B. napus* (Radke et al., 1988).

Other genes isolated from *B. campestris* are *BcNAI* (Kridil et al., 1991). ACP gene ie *1Bcs and Bce4.* Coding region an regulatory for these genes have been isolated. They have been used to design expression cassettes for testing in transgenic plants. Each has 5' and 3' non coding sequence which contain regulatory signals for tissue specific and maximum expression. A reporter gene B. glucourinidase (GUS) is used to examine function of each expression cassettes.

Napin 5' 1.75 Kb	S Bg Px	Napin 3' 1.26 Kb

Fig. 8.2: Gene expression cassette for foreign gene expression in developing repeseed. 5' promoter and 3' terminator. Letters between 5' and 3' refer to unique restriction sites of inserting foreign genes.

S=Sall, Bg=Bg12, P=Pst1, X=Xhol.

Bce4 napin cassettes show higher average levels of expression as compared to ACP. It is also seen that promoters are regulating GUS expression in tissue specific manner. Higher expression of GUS in seed coat than embryo. Choice of promoter may be an important factor for maximum phenotypic expression.

Dates show that TAG composition is determined by pool of fatty acyl Co.A available during TAG synthesis. Therefore, to modify TAG in transgenic oil seeds, one has to modify make up of acyl Co. A pools available to the TAG synthesizing enzymes.

The expression cassettes can be utilized for expressing various genes in transgenic rape seed that is genes involved in lipid biosynthesis but basic lipid metabolism in other tissues is not affected. Similar genetic engineering of novel oil composition in other oil seeds like soyabean, maize can be expected in future.

9

Recommendations and Priorities for Future Research

Traditional plant breeding and the ongoing process of improvement o agricultural practices have allowed enormous progress in recent decades ir the control of plant pathogens. Nonetheless, although this may not be apparent to the general public in North America and Europe, viral disease cause serious crop losses, not only in tropical areas, but also in temperate regions. For instance, in Asia and Africa viral diseases have considerable impact on rice productivity, and the spread of virulent strains of plum pox potyvirus in Europe will seriously compromise production of plums and apricots if resistance genes are not developed. It is thus quite appropriate tha any new means for creating virus resistance genes should be thoroughly explored for potential deployment. Considering the novelty of transgenic virus-resistant plants, it is also reasonable that the principle of precaution be applied, and that potential risks be evaluated, so that the decision to deploy on a large scale can be firmly based on an overall risk/benefit analysis.

Most of the potential risk factors associated with transgenic plants rendered virus-resistant by expression of viral sequences are in fact similar o identical to events that occur when non-transgenic plants are cultivated and are infected by viruses. There was a strong consensus at the workshop tha for this reason the correct baseline for risk assessment is the current situatio in the absence of transgenic plants. For instance, virus recombination is well known feature of virus evolution. However, only in recent years ha sequencing information made it clear that recombination has worked o viruses at all taxonomic levels, and this has led to increased interest in th detection of recombination events and in the analysis of the effects o recombination. It has been shown that recombination can occur betwee viral sequences expressed from a transgene and a infecting virus. Thus, the appropriate risk assessment question is whether recombination in transgenic plants presents risks in addition to those due to recombination in non-transgenic ones. Similarly, the baseline for potential risks associated with heterologous encapsidation is represented by what already occurs in plants infected with more than one virus, which is a frequent occurrence, and potential risks associated with transmission of virus-derived resistance genes to wild plant species should be compared to the due to transmission of natural virus resistance genes.

It has on occasion been maintained that laboratory studies are of little or no relevance to touch only situation where risk is pertinent, the open field. The results presented at this workshop abundantly demonstrated that laboratory studies are very valuable indeed. For instance, laboratory studies on the frequency of recombination in transgenic plants make it possible to design genes whose transcripts will be less apt to recombine with an infecting virus. Similarly, laboratory studies have made it possible to modify coat protein genes to reduce potential risks associated with heterologous encapsidation. More generally, laboratory studies concerning the mechanisms underlying either resistance or potential factors of risk will make it possible to carry out more pertinent studies in the field.

PRIORITIES FOR FUTURE RESEARCH

1. In studies of recombination in transgenic plants, now that it is clear that this does occur in plants expressing sequences from several types of viruses, it will be appropriate to focus on two types of questions:

2. One is a question of baseline information: is the frequency of recombination in transgenic plants greater than that in plants infected with two or more viruses?

3. The other is more a question of consequences: since recombination does occur, what is the outcome? Are there changes in host response (symptomatology), of changes in host range? Concerning symptomatology and host range, we already have some elements of response. An area in which we have much less information is: if recombinant strains do appear, how well will they compete with the parental strains? It is reasonable to predict that most recombinants will be at a disadvantage, but how general is this? In particular, if recombination leads to changes in virus host range, this would be an instance where recombinants would be expected to have a clear competitive advantage. It was pointed out that our understanding of the determinants of host range are still quite rudimentary.

4. Among virus-virus interactions, heterologous encapsidation is beginning to be well understood, but understanding is more rudimentary concerning the potential for synergistic effects, or concerning the effects of infection of transgenic plants with more than one virus. Here further studies are also clearly merited.

5. Potential risks associated with CMV satRNA genes make their future commercial use less likely, even though no adverse effects were reported when they were used on a large scale in the People's Republic of China. However, interesting questions remain concerning factors controlling the epidemiology of satRNAs occurring in nature, which could be addressed by a molecular epidemiology approach.

6. There is also a clear need for research on effects at the ecosystem level in at least two domains.

7. Since it has been shown that viral resistance genes can be transmitted from certain crops to wild relatives, can this confer an advantage to the wild species, and even enhance their weediness? Although as mentioned above this question can be raised with plants with conventional resistance genes, transgenic plants could be a particularly useful tool in such studies. This is an area where currently very little is known.

8. Molecular techniques for fine-scale studies of the composition of virus populations have come of age. It would thus be timely to apply these techniques to situations in which transgenic plants are cultivated, in order to evaluate possible effects on the composition of virus populations. For instance, it would be of great interest to compare, in the field, the prevalence of recombinant genomes within viral population in the presence and absence of transgenic plants. Countries where large-scale commercial releases of plants expressing viral sequences are current, such as the People's Republic of China and the US, provide particularly interesting areas for study.

There was also considerable discussion of monitoring associated with releases of plants expressing viral genes. Although the molecular epidemiology studies mentioned above are a form of monitoring, this clearly cannot be carried out in association with the majority of field tests. If some form of observational monitoring is necessary, how should this be carried out and by whom? If agricultural workers themselves will have to be the front line of any large-scale monitoring, where does it go from there? In many countries there is a worrisome and regrettable tendency to train fewer plant pathologists and particularly those working in the field. Classical field pathologists will be the logical relay in any monitoring scheme.

10

Potential Ecological Impact of Transgenic Virus-Resistance in Plants

Transgenic plants are considered a potential risk if they contain a train that confers a significant fitness advantage in natural situations. In order to set endpoints for "ecological impact", our aim should be the "*conservation of all plant and animal species, including those which have developed together with man, in survivable populations, in their natural communities*". This aim of human society is also agreed on by most scientists and should also apply to new evolutionary developments. One of the objectives of the United Nations Environment Programme is the conservation and sustainable use of biological diversity. Effort should be taken to establish or maintain the means to regulate, manage or control the risks associated with the use and release of living genetically modified organisms, which are likely to have adverse environmental impacts that could affect the conservation and sustainable use of biological diversity. Wild relatives of domesticated or cultivated species are important objects of conservation.

Large-scale cultivation of plants expressing viral genes developed following ecological questions:

1. Where are sexually compatible cultivar, weed and wild plant populations of the virus-influenced plant?
2. Where is the virus (naturally present and how long has the virus been there?

3. Which populations might the virus invade in the future?

4. How much is the virus (or the potential recombinant and heterologous encapsidated virus relative) able to decrease the fitness of susceptible plants?

5. What is the potential velocity of gene flow to and gene introgression into weed and wild plant populations?

For example, beet (*Beta vulgaris*) is a crop of particular concern regarding invasiveness because it commonly hybridises with annual forms of wild beet. There is still a significant lack of knowledge about the potential ecological impact of transgenic plants expressing viral sequences. Several disadvantages, but also some advantages, of the spreading of virus resistance genes to non-target plant populations are possible. Virus resistance would be most likely to cause "ecological release" in wild populations if the disease is widespread and has significant effects on plant fitness. In the future, plant breeding activities will develop improved transgenic high-performance cultivars. An ecological inventory and monitoring of wild plant habitats would help to assess the risks of transgenic gene introgression.

AREAS FOR THE ASSESSMENT OF THE ECOLOGICAL IMPACT

Transgenic plants encoding viral sequences will primarily behave in nature like their non-transgenic equivalents or their relatives. Ecological interactions must therefore be viewed in comparison to know naturally existing phenomena, e.g. gene flow, even if there is still a lack in basic scientific knowledge.

Agricultural and natural habitats

Where are locations of sexually compatible cultivar, weed, and wild plant populations of the virus-influenced plant species?

For instance, the distribution of cultivated beet, *Beta vulgaris*, is ecologically limited, due to its low competitiveness and susceptibility to a number of plant pathogens and herbivores. Therefore, an 'escape' of transgenic attributes of sugar beet is more likely via gene flow into populations of beet relatives than the escape of transgenic sugar beet itself. It is important to emphasise that all cultivated and wild varieties of beet (sugar beet, swiss chard, red beet or fodder beet) belong to the same species. This implies that there is no barrier for outcrossing of fixed genetic qualities within this species (Bartsch and Pohl-Orf, 1996). Meanwhile, beets have reached a broad world-wide distribution. With the introduction of new genes that do not normally occur in this species, either into cultivated plants or into crossable relatives, the evolutionary principle of mutation and selection is accelerated in a way that can not be expected under natural conditions, Thus it is very important to consider, case-by-case, the risks that such a manipulation could entail (Crawley, 1993; Dale, 1993; Parker and Bartsch, 1996).

Virus location and virus-host co-evolution

Where is the virus (naturally) present and how long has the virus been there?

The trait expressed by the transgenic beet in our studies (Mannerlof *et al.*, 1996), is resistance to rhizomania caused by beet necrotic yellow vein furovirus (**BNYVV**), a disease which has spread through the sugar beet fields of Europe, **USA**, Japan, and China (Cooper and Asher, 1988; Asher, 1993). The virus is transmitted via the fungus *Polymyxa betae* Keskin (Tamada, 1975; Abe and Tamada, 1986). The disease leads to decreased sugar beet yields, and a loss of up to 30% sugar content (Giunchedi *et al.*, 1987). In some places in Italy where the disease is very severe, sugar beet production has been totally curtailed (Faccioli and Giunchedi, 1974). Whole regions of the USA have been taken out of sugar beet production because of rhizomania.

Virus Epidemiology

Which populations might be invaded by the virus or the new genes in the future?

Little is known about rhizomania in wild beet habitats and about its ecological impact on wild beet fitness (Bartsch et al., 1996). However, the fact that in conventional plant breeding, wild beet germplasm can be used to achieve (partia) tolerance against against the virus (Whitney 1989), could be a hint for an ecological impact of rhizomania resistance.

Fitness Impact

How much is the virus (or the potential recombinant and heterologous encapsidated virus relative) able to decrease the fitness of susceptible plant? Out investigations simulate an outcrossing of the genetically engineered attributes to the sugar beet relative, swiss chard, as a model, and measure the performance of the hybrids under conditions in which BNYVV-resistance might be advantageous (Table 10.1). No heteroencapsidation or other non-intended side effect are reported in beets expressing the coat protein of BNYVV (Konig *et al.*, 1995).

Table 10.1: Performance of transgenic plants in comparison to the non-transgenic control *Beta vulgaris* at high BNYVV infestation level

Parameters	sugar beet hybrids	Sugar beet x spinach beet hybrids
Seedling emergence	1.11[1]	n.d.
Biomass production	1.61*	1.18*
Biomass production under weed competition	1.61*	1.23*
Bolting resistance+	not detected	5.01*+
Overwintering	1.00 (1995/96 -strong winter)	1.16 (1996/97 - 'normal' winter)
Final rate of increase (seeds produced per seeds sown)		

**statistically significant effect, + effect probably due to somaclonal variation).*

Plant Population Genetics

What is the potential velocity of gene flow to and gene introgression into weed and wild plant population? For example, *Beta vulgaris* is a crop of particular concern regarding invasiveness, because it commonly hybridizes with annual forms of wild beet. With respect to recent investigations (Boundry et al., 1993; Santoni and Berville, 1992) where markers of cultivated beets have been found in wild beet populations, we conclude that there will be exchange and introgression of the new genetic information in wild beet populations (Bartsch and Schmidt, 1997). If there is a release of transgenic sugar beets on a large scale, we have to expect dispersal of the new attributes.

ASSESSMENT

There is still a significant lack of knowledge about the potential ecological impact of transgenic plants expressing viral sequences. Several disadvantages, but also some advantages, of the spreading of virus resistance genes to non-target plant populations are possible. Virus resistance would be most likely to cause "ecological release" in wild populations if the disease is widespread and has significant effects on plant fitness. Ecological implications due to the introduction and spreading of transgenic hybrids might therefore only be observable in natural beet habitats with high level of BNYVV, where susceptible *Beta vulgaris* genotypes grow. Under these circumstances, introgression of transgenic resistance genes would surely alter the natural genetic constitution of wild populations.

In the future, plant breeding activities will develop improved transgenic high performance cultivars. An ecological inventory and monitoring of wild plant habitats would help to assess the risks of transgenic gene introgression.

Risk Assessment of Gene Flow Associated with the Release of Virus-Resistant Transgenic Crop Plants

The concept of pathogen-derived (Sanford and Johnston, 1985) has been used extensively and successfully to engineer virus resistance in numerous crops. Coat protein (CP) genes have been used, more often than any other viral-derived genes to engineer virus-resistant-agronomically important transgenic crop plants (Fuchs *et al.*, 1997).

An important environmental concern associated with the large scale cultivation of virus-resistant transgenic crop plants relates to the possible transfer of virus-derived transgenes into wild relatives through pollen flow. Marc Fuchs and Dennis Gonsalves addressed this issue by using two squash (*Cucurbita pepo ssp. ovifera*) species: The transgenic line CZW-3 (*C. pepo var. ovifera*) containing the coat protein (CP) genes of cucumber mosaic cucumovirus (CMV), zucchini yellow mosaic potyvirus (ZYMV), and watermelon mosaic virus 2 potyvirus (WMV 2) and its wild relative *C. pepo* var. *texana*. Viral transgene flow was evaluated in the field by studying the successive rate of introgression of three CP genes from F1 hybrids (= *C. texana* x CZW-3) and their backcrossed progeny (BC 1 = *C. texana* x F1 hybrids) into *C. texana* squash populations. Field plots have been maintained under low or high disease pressure by deliberately introducing CMV, ZYMV, and WMV 2 in the latter plots. Introgression occurred under low but not high disease pressure. To assess fitness, test plants (*C. texana,* CZW-3, F1 hybrids and BC 1 progeny) were evaluated fro virus resistance, number of fruit produced, and number of viable seeds. Under high disease pressure all *C. texana* plants became severely infected and produced less than one fruit per plant and no viable seeds. Hybrids of *C. texana* that contained the CP genes, on the other hand, exhibited virus resistance and produced a fair number of fruits with viable seeds.

11

Alarming Issues Related to Transgenic Plants

Transgenic crops are now being cultivated in as many as 13 countries, including five developing nations with a total area of 44.2 million hectares. During the first wave of the transgenic revolution, emphasis was laid on the so-called "input traits" that confer agronomic advantages like resistance to pests and diseases and tolerance against herbicides. About 30 per cent of the total yield is lost every year due to weeds, diseases and pests. More food would be available for human consumption if the extent of this damage could be minimised. In fact, the development of pest resistant cultivars has been a major thrust area in our classical breeding programmes.

ADOPTION OF TRANSGENIC SEEDS: PROFIT OR LOSS

Genetically modified (GM) seeds and derived foods have been the subject of a fierce debate currently raging in India, with issues such as health and ecological safety. The seeds, based upon their promise of improved productivity, have the potential to eliminate world hunger. The biotechnologists and ecologists have to examine the performance of these first generation seeds, modified for improved tolerance to herbicides, resistance to pests and viruses, with particular reference to farmers.

USA is the largest grower of transgenic seeds, close to 29 million hectares under cultivation (Financial Express, 06-12-2000). These are surprising results. A study by the US Department of Agriculture's Economic Research Service has showed that yields were not significantly different in engineered versus non-engineered crops in 12 out of 18 crop/region combinations, HT cotton showed no significant yield increase in either region where it was surveyed. Another study, which examined more than 8,000 field trials, found that a particular HT soybean brand produced fewer bushels than similar conventionally bred varieties.

HT soybeans have resulted savings in herbicides use, as the use of other herbicides has declined significantly of offset the increased use of specific herbicides such as glyphosate and glufosinate. However for Bt corn, cost related to insecticide use have actually increased due to the tendency of Bt crops of lose their pest-resistance traits in the medium-long term. Here again, their performance is mixed. So, farmers, in actually, have not achieved either higher yields or the promised costs savings which they expected, and which drove them to adopt seeds in spite of them being more expensive.

Who has benefited?

Studies attempting to analyses the breakdown of profitability of GM crops between biotechnology/seed firms and farmers have indicated great news for biotech companies and not so favourable results for farmers. For HT crops, companies profit enormously from their combined pricing strategy and for Bt cotton, companies gained 74 per cent of the total rents against farmers' shares of 21 per cent. Poor yield performance, increased costs for certain crops and profits accruing to big corporates-not really a savoir for Indian farmers, are they?

The minor yield increases documented so for have been due to GM seeds enabling farmers to practice precision agriculture and narrow row farming, all very difficult to achieve when the majority of the farming continues to be carried out on small, fragmented land holdings with few technological in puts. Only the very rich farmers would be able afford these more expensive seeds. Moreover, these crops have led to increased flexibility in agricultural practices, such as simplified weed control, conservation tillage, broad spectrum control etc, all of which have led to reduced labour requirements. This convenience effect can be expected to have a significantly adverse economic impact on labour intensive developing country agricultural systems, such as ours (India's agricultural sector employs 64 per cent of the workforce).

Additionally, the intellectual property rights system intrinsic to the World Trade Organization is heavily biased against developing countries (Financial Express, 06-12-2000). Not only does it provide MNCs the right to size and patent genetic resources without any/adequate compensation, but it also prevents farmers from saving and reusing the modified seeds. This is a very serious concern as traditionally. Indian farmers have relied upon saved seeds for their cropping activities the following year. Moreover, being forced to come back to seed giants year after year, for seed purchases would be financially unviable for them, leading to their increased vulnerability.

GM seeds do have the potential to revolutionised world food production systems, but, are they necessarily to our advantage? The role of the private sector must be evaluated in this regard. Globally biotech companies have spun off their agriculture "businesses from pharma ventures and entered into collaborations with grain/food processors at the lower end and food retailers at the upper end of the supply chain. Retailers would research consumer tests and seed companies would then develop seeds with the required output traits, simplifying the task of the food processors. Such a strategy is a win-win situation for the food industry but could contribute towards the increasing marginalisation of the farmer's role in food production.

Apart from the unresolved controversy pertaining to their health risks, there are also genuine environmental concerns associated with GM crops which India would do well to consider. The dramatic effects to rotations and intercropping on crop health and productivity have been confirmed by scientific research. Because of the convenience they afford to growers on account of their producer-friendly traits. GM seeds could encourage monoculture cropping contributing to further decline in land productivity and genetic diversity.

The Joint Parliamentary Committee approval for the draft Protection of Plant Varieties and Farmers Rights Bill, India's proposed amendment of the Seeds Act, 1966, to restrict the entry and used of GM Commission, to make labeling of GM seeds mandatory, are all steps in the right direction provided GM seeds need to be adopted by Indian agriculture. However, the evidence that has emerged so far is to the contrary. The next generation of transgenic seeds would give rise to crops modified to adapt to extreme soil and weather conditions. Of course, they would reduce stress on natural resources and increase productivity. However, the problem in India is not about inadequate production, but inadequate storage handling and distribution facilities leading to tremendous wastage.

Indian would do well to examine issues like farmers rights, lack of infrastructure to identify, segregate and/or prevent the entry of GM seeds, intellectual property and its implications. Most important, there is a need to prevent wastage and losses plaguing the system.

LEGAL RIGHTS IN THE NEW BIOTECHNOLOGY

New biotechnology has developed many issues concerning the legal rights of newly introduced biotechnologically created new products for farmers, plant and animal breeders, and consumers. Various forms of legal protection are available, of varying degrees of importance and applicability, depending on the specific process and nature of the protection sought. Copyright, Plant Breeders' Rights and Patents are the main areas of the law that are relevant.

What is a Patent?

A patent is a legal monopoly granted by a national government to an invenior, which allows him/her to exploit his/her invention by preventing anyone else from copying (*infringing*) it for a limited number of years (normally 20) in return for the full disclosure of the invention to the public and payment of sizeable official fees to the government. The idea of patents, historically, was to give inventors an incentive to make public new discoveries for the benefit of the State.

An invention must fulfill following criteria:

1. It must be *new.*
2. It must not already known.
3. It must be inventive over.
4. It must be capable of industrial application.
5. It must not be in one of the categories of matter, which are considered incapable of being protected by patents.

Animal and plant varieties as such are one of the these categories. But microbiological procèsses or the product of such a process are patentable, and inventors can therefore patent new biotechnological advances in theory and gain monopoly rights to exploit them.

Impact on Farmers and Consumers

This has led to a number of scare stories, mostly put about by people with no direct knowledge of patents or how they are obtained. Farmers, plants and animal breeders are being told that the biotechnological multinational companies will be able to gain patents that could, for instance, prevent them from freely developing new strains, or force farmers to pay substantial royalties on a new, patented product.

This is a legitimate area of concern, but should be kept firmly in proportion. Although many patents have now been granted for biotechnological processes, many of these are in pharmaceutical or medical fields, which only indirectly concern farmers and breeders.

Various groups have traditionally opposed to patents as such. One should not forget that the main value, to the consumer, of the patent system lies in the encouragement it gives to inventors to bring new products onto the market-place. Many domestic products from which the consumer has benefited products are the subject of patent protection. The inventor has benefited as well, but not at the consumer's expense.

The *Working* patent made its inventor, a millionaire from the royalties on it paid to him; but the manufacturer paid these royalties. That proportion of the cost to the consumer of the *Workmate* which covers the royalty is negligible when set against, say, the proportion covering advertising costs. Furthermore, having brought the product, it is fundamental to English Law that its owner can enjoy the use of it without further payment.

Ethical and Practical Problems

The rights should not be granted on living matter. Such concern may arise from religious conviction, or from the realisation that we have no moral right to unreasonably exploit our world.

The main difficulty lies in defining the *invention* in a way that is legally acceptable. The hole basis of the patents system is that monopoly rights are granted for the full disclosure of what is new about the invention to the public. But how do you define a biotechnological advance clearly enough to be acceptable, and ultimately to stand up in Court before a hostile barrister? it is possible to adequately define an invention in terms of the *bits* that make up its DNA structures. But even the very simplest life forms take many pages to describe in these terms.

A problem which is easier to overcome, but cannot be ignored, is that worthwhile patent protection on an appropriate international scale is not cheap. Even the simplest patent application with the necessary professional advice, in the UK alone, may cost around 600. For a complicated, difficult, biotechnological patent costs may be ten times this. Since there is no truly international system of patent protection (although a form exists in Western Europe) patents will need to be applied for on a country by country basis. Where translation into the country's own language is involved, costs would typically be at least 3000 per country. Furthermore, this money must be spent *up-front*, normally within 12 months of the home application being made otherwise the chance of protection abroad is lost entirely.

The particular relevance to the biotechnology industry is the timescale involved in developing a patent. A patent can only be enforced against potential infringers once it is granted: but this may take up to live years from the date of the application. Yet, that same patent application should be published, and therefore open to public inspection, after only 18 months. Companies are understandably reluctant to allow their competitors access to their latest developments up to four years before they could stop the competitor taking unfair commercial advantage.

Given these drawbacks, it is perhaps not surprising that many of the big companies at the sharp end of the bio-revolution have chosen to ignore the patents system entirely. They rely instead, during the development of the product, on rigidly enforced *Confidentiality Agreements,* which bind their present and former employees not to disclose what they are working on outside the company; and in many cases also on closely controlled *cell* structures within the company itself so that only the minimum number of people ever have access to any one product. Once the product comes onto the market-place, the company relies on its commercial muscle, backed up by an effective and legally protected trademark, to promote *brand awareness* and hence to establish a dominant position in the market that potential competitors are effectively deterred due to the vast sums of money that would be required to challenge such market dominance.

Multinational companies responsible only to their own shareholders and operating behind a well-established veil of secrecy may have a more direct and far reaching effect on the farmers and breeders who seek to develop and exploit new biotechnological advances than any legal monopoly nights.

TRANSGENIC ARE NOT TERMINATOR CROPS

The most debated topic in agriculture and biotechnology is genetically modified (GM) crops, that is, transgenics. Genetic engineering (GE) technique allows the transfer of a gene of interest from one species to another. It is nothing but the extension of conventional plant breeding. This technology helps the breeders and molecular biologists to introduce the only gene of interest with more selective modification and represents a significant advance.

Terminator technology is a technique of genetic engineering by which the plant is altered or modified in such a way that it produces sterile seeds, which will not germinate in the next generation. Normally, seeds have the ability to germinate generation after generation. In terminator seed, a toxic gene becomes active and is forced to die after one generation of survival. In terminator technology, three gene constructs are introduced through GE techniques to express the desired character, sterility.

The three gene constructs or stretches of DNA are referred as gene X (lethal gene), Y and Z. This entire system is under the control of an external inducer, that is, tetracycline. If only the lethal gene is inserted, then the first generation itself will not germinate. Initially, the lethal gene is prevented from being expressed for seed multiplication and thereafter, gene Y helps to excise out the 'blocking sequence' to make the gene X (ledder the control of a different gene, Z. The lethal gene becomes functional only at a later stage of seed development and allows the gene to produce toxin that kills the embryo. Ultimately, the seed harvested by the farmer lacks germination ability but according to paten claim, the seed or fruit can be used as food.

Terminator technology has been developed to serve two main purposes. To force the farmers to buy seed each year for raising the next crop and to prevent the piracy of the introduce special genes (that is biotic and abiotic resistance or nutritional characteristics) by researchers and multinational companies.

The Hypothetical Risks

There are following alarming issues to be answered:

1. How safe are these produced seeds from the terminator seed for human consumption?
2. And how safe to the environment as the activation of terminator gene requires pre-soaking of the seed in tetracycline?
3. Will these seeds respond differentially to environmental changes or have a devastating effect on bio-diversity?

The beneficial side is that this technology has the ability to prevent escape of transgenes to bio-diversity.

Biotechnology is used to harness the problems that are difficult to overcome by traditional plant breeding techniques. GE techniques are primarily applied to resist biotic and abiotic stress, higher productivity, better seed and nutritional quality. This technology is helpful in resistance breeding, introducing gene, irrespective of their presence in nature. 'Bt' gene from the soil, bacterium - *Bacillus thuringiensis* is introduced to crop plants for controlling lepidopteron insects in man crop plants.

In cotton, *Bollgard* a transgenic variety resistant to cotton genic variety resistant to cotton bollworm, is already in commercial cultivation and gained popularity. The herbicides are used to kill weeds but, unfortunately, they kill crop plants too. Therefore, the transgenic crop plant tolerant to herbicide are an important requirement to control weeds through herbicides. Development of bio-pesticides is an emerging area of molecular breeding, with positive contribution to integrated pest management. GE techniques are used to improve the virulence of natural enemies of pests.

The major component of a human diet is oil and fat, in which saturated fatty acid compounds are present, which are hazardous to health. GE techniques have proved their ability to reduce the level of saturated fatty acid components in soybean and castor. Balancing of human nutrition may be carried out by trangenics to supply balanced source of amino acid. Curing human disease is becoming possible through the use of transgenic plants. Recently, it was reported that transgenic potato and tobacco can cure cholera. A transgenic potato can also prevent diabetes. European researchers developed a vitamin rich (carotenoids) tomato, which eventually, could prevent heart and cancer disease.

Biotechnology has the potential to help society solve serious problems, but the new technology has to be handled very cautiously to protect the flora and fauna from unintended consequences. The fear is that the transgenic crops will become weeds and that novel genes may be transferred to wild populations, leading to *superweeds*. This is not that easy, as it depends on the nature of pollination and many other factors. In case of self-pollinated crop, the extent is less than five per cent, whereas in cross-pollinated crop it is up to 50 per cent and for wild and weedy relatives, the gene flow is very less (less than 1 per cent). The effect of gene flow in case of weedy species is not dangerous enough to cause extinction of biodiversity. Hence, the fear of loss of entire bio-diversity needs further critical examination.

Though the inception of genetic engineering in agriculture has to face scientific and technology challenges, economic, regulatory and social concerns, it will emerge an important tool in solving the problems of food production healthcare, bio-diversity and environment conservation. The present world population is around billion by 2020, it may becomes billion and out of which 85 per cent would be from the development countries. It is time to increase India's agricultural productivity which is still far below the world average in most food crops.

The Green Revolution was possible due to the genetic manipulation of the plants by plant breeding. After the Green Revolution, most crops are suffering from yield plateauing and further boost may be possible from molecular breeding. To meet the challenges ahead, funds must be diverted towards new technologies to improve agricultural productivity. Biotechnology may be able solve some of the problems associated with lower productivity, and could play a major role in agricultural production. A survey conducted among Indian farmers shows that they are highly innovative and interested in adopting any new technology. But it should be made available to them.

> *A misconception still prevails among farmers about transgenic and terminator technology. They confuse one for the other and think that any transgenic crop will lead to sterility in the next generation crop. In both cases, the product is developed by using the same techniques of genetic engineering but the products are developed for different purposes.*

The governments of developing countries have identified biotechnology as a promising and key research area with widespread application in diverse field. The developed nations benefit from biotech products because, of both private and public sector collaboration. Though there are long term risks and hazards associated with it, this new technology should be used to complement, not replace, traditional plant breeding methods to enhance productivity and quality.

GOVERNMENT ALERTED: DID GM FOODS REACH INDIA?

Experts say unlabelled food was distributed among Orissa cyclone victims (Economic Times, December, 14, 2000). After the brouhaha over cheap Chinese imports, the focus has now shifted to genetically modified foods (GM foods). Stung by the criticism that genetically engineered foods may have *unknowingly* food their way into India as part of US aid and relief to the Orissa flood victims, the government has begun the process of looking into the issue. As a result, the commerce ministry has been directed to prepare a note on GM foods, which may figure as part of the deliberations on Agreement on Agriculture (AoA), scheduled for review in March.

The good and ill effect of GM foods have been debated endlessly without any sign of a consensus emerging either globally or at home. If there is a strong enough lobby opposing even the setting up of a working group on biotechnology at WTO-issues that make it to the working group level normally become acceptable as heads of negotiation under the WTO-there is a strong group which believes that India should not oppose the US, Canadian and Japanese move. So also is the issue of GM foods.

However, recent allegations by experts like Dr Vandana Shiva, Director, Research Foundation for Science, Technology and Ecology, the GM foods have clandestinely found their way into the country, have once again raised the spectre of Frankenstein Foods. It is Dr Shiva's contention that the US has used even a mega disaster like the Orissa cyclone, which killed 35,000 people, to dump *unlabelled* soya and corn on a predominantly rice-eating population. Medical professionals at AILMS too have alleged similar things, which has resulted in forcing the government to examine the issue.

Part of the allegations stem from the fact that about two-third of the food ($.5 million out of the $6.5 million US aid and relief for Orissa), which was imported for the victims was *not labeled*, -at a time when UK, EU and several other countries have closed their doors to genetically modified foods – is being interpreted as these foods being "rejects" which found their way into India.

Similar allegations also surround the genetically engineered Vitamin A rice. Here, it is being alleged that government and quasi government agencies are being roped in by seed companies under the pretext of blindness alleviation to popularize and market the new product.

While, these are issues, which are likely to be settled only by scientists, doctors and biotechnologists, some facts remain. At the peak of the GM food wave, seed companies like Monsanto invested heavily in biotechnology R&D. Modified seeds are the end result. These companies are undoubtedly on the prowl, looking for new markets to reap a return on their investments.

It is also true that propaganda against Frankenstein Foods (GM foods) has reached feverish pitch in the West. Everybody starting from Prince Charles to Paul Mcartney (ex-Beatle) have spoken against GM foods.

The EU, which is the biggest importer of food stuffs, has indicated that it would like to renegotiate the Sanitary and Phytosanitary Systems Agreement to permit the use of trade restrictions on grounds of consumer preference.

As a result of EU's position on GM Foods, US has fallen out as trading partner. The last two years, particularly 1999, have seen a reversal of GM food's fortunes, with at least half a dozen companies and retail chains publicly promising to shun GM foods. Some of the big names include Mark and Spencers, Sainsbury, Carrefour, Irish Superquinn, Tesco, Unilever, Nestle, Cadbury, McDonalds and Blake Brothers in UK and EU. Iceland has taken the whole debate a step forward by becoming the first company to ban GM Foods in its feedstock.

The US government has consistently opposed the labeling of GM foods. At Seattle, US President Bill Clinton made no bones about the US government's support to biotechnology and bio-foods. US commitment to GM Foods stems from the fact that it is the foremost country which has put large tracts of land under GM Food cultivation, followed by Canada, Australia, Argentina and Japan, to a small extent. Its companies are the largest investors in novel foods.

The emerging opinion is that if at all India does decide to provide access to genetically engineered foods, that it should do so consciously rather than open its doors to clandestine entry.

In that case, India would have to join hands with other countries to push for labeling of at least those GM-foods, This article represents a review of some of the ethical dilemmas that have arisen as a result of the development and deployment of transgenic crop plants. The potential for transgenic crops to alleviate human hunger and the possible effects on human health are discussed. Risks and benefits to the environment resulting from genetic engineering of crops for resistance to biotic and abiotic stresses are considered in addition to effects on biodiversity. The socio-economic impacts and distribution of benefits from transgenic technologies are reviewed. An almost unprecedented amount of discussion has been stimulated on the merits and demerits of genetic engineering of crop plants, and has divided both the public and scientific communities. The arguments for and against transgenics are invariably based on visions of the new technology from widely different ethical perspectives. Fundamental issues of man's relationship with nature and the environment, and theological matters are also addressed.

TRANSGENIC PLANTS: BIOETHICAL AND MORAL CONCERNS

Various terms have been given for transgenic technology as, *Cashing in on Hunger*, *Demon Seeds*, *Terminator Technology* and *Frankenstein Foods.* However, it is difficult for anyone to appreciate any underlying truths, should they even exist for transgenic plants. The cultivation of transgenic crops could, through intensification of agriculture, contribute to increased agricultural production and an alleviation of human hunger, while promoting environmental conservation.

Bioethics

Bioethics especially includes medical and environmental ethics. Bioethics considers the ethical issues raised in biology and medicine, especially those raised by human activity in society and the environment using biotechnology. Bioethics considers issues affecting all living organisms and the environment, from individual creature to the level of the biosphere in complexity. All living organisms are biological beings, and share a common and intertwined biological heritage. The term bioethics reminds us of the combination of biology and ethics, topics that are intertwined.

Agricultural systems include economic, environmental, and human interactions. To resolve the issues, and develop ideals or principles to help us do so, we must involve anthropology, sociology, biology religion, psychology, philosophy, and economics; we must combine the scientific rigor of biological data, with the values of religion and philosophy to develop a world view. Bioethics is, therefore, challenged to be multisided and thoughtful approach to decision making so that it may be relevant to all aspects to human life.

Some academics have tried to define more precisely what bioethics is, and basic principles. There are two basic approaches, one being descriptive and the other being prescriptive. One describes how people makes decisions, and the other suggests the process that can be used to make decisions. However, bioethics is not to prescribe the correct *answer*, only the process that is used to decide. To make good choices, and choices that we can live with, improving our life and society, is certainly a good thing. However, what is good for one person may not be good for the broader society, and the global nature of agricultural economics and environmental impact, makes us think of the global arena. The choices that need to be made in the modern biotechnological and genetic age extend from before conception to after death-all of life.

The bioethical concerns are already one factor affecting the implementation of biotechnology in agriculture, because they are part of the "equation" that people use when deciding whether to accept the products. It is in the interest of the whole of society that persons make good decisions considering the total current and future effect of their choices, and scientists have a role in education to enable this.

Ethical Concerns

The fundamental ethical concerns of biotechnology under the heading *intrinsic* have been pointed out by (Straughan, 1995). He raised the issues of theology, naturalness and respect for nature. Deeply held beliefs, that clearly separate right from wrong, exist for many and have to some extent been championed by Prince Charles (1998). They claimed that GE takes mankind *into realms that belong to God and God alone*.

It may be stated that there is a moral obligation to supply transgenic technology to areas where human hunger could be lessened, but there will need to be a case-by-case assessment done, and transgenics appear not to represent a panacea. Transgenics are unlikely to become more popular however, if the business ethic is seen to prevail over human welfare and the environmental ethic.

Transgenic plants embraces various major issues in philosophy; the meaning of words, free will, the concepts of right and wrong and justice, and ultimately the meaning of life itself. It has become a battleground involving scientists, commerce, politicians, journalists, lobby groups and the public. The arguments are about values of transgenic plants. These are neither absolute nor universal. The controversies surrounding transgenic crops have to a large extent polarised society into the proponents and opponents. These plants once seemingly trustworthy and ethically sound.

The risk and benefits of transgenic plants associated with man and the environment are varied. The balance of the distribution of benefits, and on the technology being good or bad in itself. The moral and ethical concerns are important factors in influencing a risk-averse public (Callahan 1996), and are a pivotal feature of the debate on transgenic crops and their products (Wadman, 1996; Geary, 1996; Newton *et al.*, 1999).

Transgenic Plants: Risks, Benefits and Impacts on Society and the Environment

Concerns about the consequences of development and deployment of transgenic crops, the risks, benefits and impacts, which are referred to as extrinsic concerns by Straughan (1995), are foremost in many discussions of the relative merits of the new technology. Following are the basic questions which require answers:

1. Do transgenic crops represent the solution, albeit possibly partial, to world hunger.

2. Do transgenic crops represent unacceptable risks to the environment and human health.

3. Do transgenic crops represent a means for improving equitable sharing of the benefits of technological advance?

These issues are inherently linked and any absolute division is artificial.

Agricultural research and its role in food production as being intrinsically good (Hardon, 1997). This view has been altered somewhat by the advent of GE (Wagner, 1996), although it is generally appreciated that new technologies by their very nature represent a challenge to existing values and systems, and stimulate change in traditional concepts of nature and human identity (Carr and Levidow, 1997). According to Borlaug (1997) the myth that agriculture is practised in a rural Eden needs to be dispelled and replaced with the reality of it being a struggle to produce food for an ever increasing population against natural forces.

Transgenics and Human Hunger

Food shortages generally occur in areas characterised by poverty, high population growth rate and political instability, if not war itself. Transgenics may reduce poverty and hunger and promotes equity in many of the world's poorer countries (Serageldin, 1997; Ortiz, 1998). It is Malthusian preoccupations, feeding a human population of ten billion in the foreseeable future, which represent the ethical justification for employing such biotechnology (Jaffe, 1994; Borlaug 1997). Food shortage may cause hunger, and ignores to some extent the reasons for poverty, inequitable distribution of food, land tenure inequity, overpopulation, poor health, poor education etc. Carr and Levidow (1997) emphasised the multi-faceted nature of hunger. Therefore, transgenics have a moral head start on other technologies.

Is GE can make a positive contribution to relief of hunger where it occurs?. The biotechnologists still have to solve this problem.

How transgenic crops might contribute to sustainable farming practices (Pretty, 1999)?. Furthermore, transgenic research is mainly conducted by chemical companies and directed towards chemically dependent crop varieties (Hubbell and Welsh, 1998) and value-added products rather than staples (Levidow and Carr, 1997). Rifkin (1998) believes that the current type of agricultural biotechnology is misdirected and promises disaster, while Simmonds (1997) believes it to offer more than it can deliver and Serageldin (1997) stresses that the *opportunities for producing transgenic varieties are endless.* Shiva (1998) believes GE will displace industries in poorer countries and jeopardise smallholder farming. Therefore, the range of opinion on the probable impact of transgenic crops is wide.

Where will the required increase in agricultural output take place?

There are basically two following options:

1. Intensification of agriculture in areas currently farmed.
2. Expansion of the area under cultivation to take in new areas, as yet uncultivated.

Using current intensive farming methods the first option promises further deterioration of already damaged environments while the second will result in the loss of delicate ecosystems, such as tropical forests and savannahs, and their associated biodiversity.

The Poor Farmers

Poor farmers unable to purchase sophisticated inputs such as transgenic seed. They are high priced. Mainly seed companies develop transgenics for those areas where poor people live. It is unlikely that the altruism of big business stretches so far, although poorer countries might represent a good test-bed for new technologies. As a direct and sole solution to human hunger it seems unlikely that transgenics will make the required impact (Pretty, 1999).

TRANSGENICS AND THE ENVIRONMENT

Concerns about the natural and agricultural environments differ according to country: there is less concern for agricultural environments in the USA, however, developing countries are more concern about it. Intensive agriculture adversely affects the environment because farmers are using chemical inputs for optimising soil nutrient conditions, seeds of varieties correspondingly responsive to such conditions, and pesticides for controlling insects, pathogens and weeds (Harvey, 1998). Pimentel (1995), Paoletti and Pimentel (1996) and Harding and Harris (1997), among others, have reviewed risks and benefits of GE in agriculture, and Johnson (1999) presented a case for exercising precaution in releasing transgenic crops into the European landscape until more is known about their effect on natural biodiversity.

Crop plants engineered to suit the environment better through incorporation of genes for tolerance to biotic and abiotic stresses (Ortiz, 1998), and thereby an ethical advance, while others regard such crops as being just as environmentally unfriendly as the technologies they are supposed to supersede (Rifkin, 1998). The immediate environment, farmland, and the surrounding, non-farmed environments could be affected by introduction of new technologies.

GE of crops for reduced fertiliser requirement through *in planta* nitrogen fixation could be beneficial through reducing the negative impact on the soil and the subsequent effects of run-off into rivers and seepage into ground water. Hubbell and Welsh (1998) documented three following classes of transgenics;

1. Transitional traits, which reduce environmental damage in the short term through substituting for an input.
2. Compatible traits, which reduce use of non-sustainable inputs without polluting the environment.
3. Sustainable traits, which are fully sustainable over time.

Objections to Development and Deployment of Transgenic Crops

Transgenic plants have developed several issues relating to the balance of associated risks and benefits. Many crops have been engineered to withstand herbicide application and there is much debate on the following issues (Kling, 1996).

1. Whether this will lead to more or less herbicide being applied to crops.
2. Which types of herbicide (in terms of environmental friendliness) will be applied.
3. The persistence and effects of herbicide residues.
4. The possibility of herbicide resistance developing in target species.
5. Genes of transgenic plants being in turn passed on to non-target relatives to create invasive herbicide resistant weeds.

Transgenic crops and their genes, through pollen dispersal, can spread (Brookes, 1998) even between species that are mainly inbreeders (Cavan *et al.*, 1998). Regarding transgene and transplastomic containment, there are no standard scenarios (Chamberlain and Stewart, 1999), but it has been established that there is a low probability of chloroplast movement from oilseed rape into wild species (Scott and Wilkinson, 1999). The effects of transgene escape on the environment are uncertain, but modern technology could limit such **genetic pollution** through, in some cases, engineering sterility into the transgenics to ensure vastly reduced gene flow into the farming and natural environments. Crops do not generally survive outside the farming environment and transgenic crops would probably be out-competed should they spread off farm.

Pest and Disease Resistance and Transgenic Technology

Engineered virus resistance could result in the evolution of new and harmful viruses (Borja *et al.* 1999; Rubio *et al.* 1999), and crops engineered to produce toxins (Bt toxins in the main) might poison non-target hosts (Concar, 1999). There are differences in outcrossing rates between transgenic *Arabidopsis* plants and mutant *Arabidopsis* expressing the same mutant allele (Bergelson *et al*. 1998).

Various targeted pest species have developed resistance to engineered genes. It occurs in the same way as they have done to naturally occurring resistance genes (Holmes, 1997). It is an example that nature fights back against the genetic engineer in much the same way as it fights back against the conventional plant breeder and many solutions to pest and disease problems represented by GE are likely to be short-lived. Pest control by using agrochemical is however inefficient and environmentally and ethically unsound (Pimentel, 1995). GE technology offers a remedy, allowing more precise targeting of pest management.

Transgenic Trees

In future, trees will be the next targets on the agenda for genetic engineers Ortiz (1998). If it happens then a new ethical and environmental concerns will emerge because of the following reasons James (1997):

1. Long-lived and little domesticated nature.
2. The problems that will ensue through possible pest adaptation.
3. Production of environmental toxins.
4. Enhanced invasiveness and transfer of transgenes.

Acording to James (1997): *man has an ethical obligation to the ecosystem and must be considered when balancing the risks and benefits of GE of trees for insect resistance versus application of toxic pesticides.*

James (1997) indicated the similarity between biocontrol and GE. Biocontrol is being concerned with the introduction of entire novel genomes, while GE with the introduction of novel genes.

Hokkanen and Lynch (1995) dealt with benefits and risks of biological control, including use of GMO. Many of the ethical issues are the same as those associated with GE, but it is interesting that biological control appears to have attracted a lot less attention from the media and the public than transgenic technology and has consequently had a much more positive press.

Transgenic crops may cause genetic diversity and its possible erosion. Hardon (1997) pointed out that plant breeding relies on genetically diverse germplasm for progress to be made and maintained, which in traditional agriculture is regarded as a common resource of great value and is freely available. Hardon (1997) considers it unethical to treat such traditional forms of agriculture as markets to be conquered by private interests (**biopiracy**).

The development of transgenic crops will enhance biodiversity by creating an increased need for exotic genes. Genetic diversity will be diminished through deployment of a narrow range of germplasm. Biodiversity, both on and off farm, is already in decline phase. It is due to current farming practices in both the developed and developing world. Whether transgenic crops will accelerate or dampen? . This remains a debatable issue. The dilemma is that the greatest biodiversity exists in the delicate ecosystems of the as yet uncultivated areas and the cropping systems of some of the world's poorest countries.

The effects of transgenics on the environment are controversial because of the difficulty in gauging the associated risks (Weil, 1996). Although all actions are potentially hazardous, there have been no problems involving transgenics to compare with those that have been encountered previously as a result of classical plant breeding.

TRANSGENICS AND HUMAN HEALTH

Plants are the major sources of the human food. A major worry of the public, aside from fundamental concerns, is that transgenic crops contain ethically sensitive genes, including, for example, antibiotic marker genes and promoter sequences derived from viruses. These genes have aroused concern to human health and may be adversely affected by consumption of transgenic crops and products derived from them (Coghlan *et al.*, 1999). This aspect has developed to the extent that restaurants have removed GM foods from their menus and schools have outlawed GM products (Ward and Hall, 1999). There has been increasing demand in supermarkets for organic food (Clover, 1999). *Is this the start of a biotechnology backlash* (Williams, 1998)?

Bengtsson (1997) maintained that as some crop varieties will be transformed many times, antibiotic resistance genes will accumulate, and it is therefore sensible to remove them as plant breeders will soon encounter difficulties in locating new, harmless antibiotic marker genes. This is an obvious fear.

The World Health Organisation (WHO) has judged antibiotic marker genes to be safe (WHO, 1994), but the outcome of their use might be hazardous if they represent a major source of resistance to a wide class of antibiotics. The fear is that these genes could spread from the plants into which they were inserted to wild plant populations and to bacterial populations that would then be advantaged in their natural environment.

Transgenic foods may be toxic or allergenic. Franck-Oberaspach and Keller (1997) reviewed the consequences of classical and biotechnological resistance breeding for food toxicology and allergenicity. They reported on many classes of actual and putative toxins and allergens, concluding that several naturally occurring defence substances found in plants are highly toxic to mammals, but also indicating that food safety can be severely influenced by natural pathogens and their products.

It is difficult to maintain good health on a purely vegetarian diet, but what if plants were made healthier through GE, would this be ethically sound from society's and the consumers' points of view? Strict vegetarians might object to gene sequences from animals being introduced into plants. Following questions lie at the heart of the debate on transgenic crops:

1. What of the benefits that have accrued for humans from GE to produce insulin and human growth hormone; are direct medical applications of GE regarded as being different from those which alter plant characteristics for an indirect benefit such as nutritional improvement?
2. Does GE of humans, animals, plants and microorganisms require different ethical considerations.

Human health already suffers as a consequence of agricultural practices. Commercial banana production, for example, requires application of large amounts of pesticides which pollute the environment, and whose residues accumulate in plantation workers. It is ethically not justifiable to produce a transgenic banana variety that allows for a reduction in pesticide application and a subsequent improvement in human health.

Socio-Economic Impact and Benefits

A large-scale farmers always favours transgenic technologies. This will cause loss to the third world markets through export substitution (Weil, 1996). For example, the artificial sweeteners cause negative effects on the sugar industry of the tropics. Similar situation is for vanilla and various oils (Mannion, 1998).

It seems that the poorer section of society is bound to lose out.

Transgenic technology has been described recently as **Terminator Technology** (Service, 1998; Crouch, 1998). It has raised substantial ethical concerns in that it provides a means of ensuring that seed cannot be saved at the end of one crop cycle for sowing at the following cycle. This would potentially put the farmer very firmly under control of the company providing the seeds and would, if deployed in such areas, preclude using home-produced seed, a common practice in many parts of the world, not just in developing countries. This puts the business ethic and humanitarian ethic in direct conflict, and yet this same technology could be harnessed to limit natural dispersal of transgenes.

OWNERSHIP OF GENES AND THE NEED FOR PATENTS

This is a further area for ethical debate. Patents and plant breeders' rights have largely corrected for this unfairness, but what of transgenes? Uncountable numbers of exotic genes have entered crop varieties through conventional crossing programmes, and it would be impossible to trace them back to their origins and compensate the owners.

King and Eyzaguirre (1999) discussed **Intellectual property rights** (IPR) for protecting indigenous resources. They pointed out that biological resources are frequently integral components of cultural systems which, in the face of technological advance, are threatened to an equal extent as the physical and biological resources. It is essential that attempt should be made for protection of biodiversity through IPR considering socio-cultural setting, specially, value systems and ownership concepts.

Genetic Resources Action International (GRAIN, 1995) suggested that there are viable alternatives to western IPRs, especially within the area of biodiversity. It maintains that IPRs, although once used to stimulate innovation and reward invention, are now used for economic expansion and market control, and particularly so following developments in genetic engineering. Moreover, GRAIN (1998) goes further in suggesting that the World Trade Organisation (WTO) is assuming an ever-larger role in intellectual property regimes through its Agreement on Trade Related Aspects of Intellectual Property Rights (TRIPs), which will have to be implemented by developing countries. It is suggested that costs will outweigh the benefits and that the outcome will be destruction of the socio-economic fabric promoting innovation in developing countries.

GMO AND HEALTH PROBLEM

Americans are far ahead in research on this subject and have been consuming GMOs for years. To date, there has not been any scientific evidence of health problems caused by consuming such products. In your report on Greenpeace's exposure of the GMOs in food products (Bangkok Post, April 11), it did not and could not substantiate how the GMOs had any ill effect on human health. Many countries who lag behind in research and production are against GMOs for obvious reasons, and so put up trade barriers. If Greenpeace is sincerely concerned about the health of people, it should have exposed the real, hazardous substances contaminating foods one consumes daily, such as insecticides in vegetables and fruits, formalin in fish and meats, hormones in chicken and pork, antibiotics in shrimp and chicken, as well as the production of narcotics that are ruining the brains of Thai children.

ANTI-GMO PROTESTORS TARGET

Demonstrations are going on in various parts of world regarding uses of GMO. Recently, semonstrators staged protests in several parts of France (Agence France Presse April 17, 2001) as part of a worldwide day of action to demand a halt to the cultivation of genetically engineered crops. Protestors gathered outside the premises of the country's gene research programme, Genoplante, in Evry (Northern France) and the activists stormed an experimental farm in Aveyron (South East France). The local authorities in the southeastern town of Cabeuil ordered a ban on all planting of genetically modified organisms (GMOs).

These protests were the part of a worldwide campaign launched by *Via Campesina*, a global network of farming, environmental and other organisations including Greenpeace and France's Cofederation Paysanne (Peasants' Confederation), whose leader Jose Bove is famous for spectacular protests against globalisation and GMOs. Jose Bove has criticised the five-year Genoplante programme, whose budget of 1.4 billion francs (0.19 billion dollars, 0.21 billion euros) is largely state-funded, for using public money to *further the interests of private companies* and assist them in *gaining ownership of living organisms* by patenting gene-altered crops.

The supporters of research into GMOs have different openion. They say that *genetic engineering could help make plants more resistant to viruses, drought and highly saline soils, and therefore boost crop yields.*

But opponents say that *GMOs resistant to herbicides and insects could pass on these properties to other species, with potentially disastrous consequences.* Multinational biotechnology firms are keen to introduce gene-changed crops to developing countries in a bid to erode growing resistance to GMOs in Europe. But Greenpeace insisted earlier this week that *third world hunger was more a political problem than an agricultural one and said gene modification could enable biotech multinationals to produce alternatives to developing country crops* such as vanilla, thus further undermining the latters' economies.

GM TIDE TURNS THE TIMES

Public Opinion Swings Towards GM Foods CropGen

Recently, in UK, a majority of people are happy to eat genetically modified foods for the first time since the issue emerged three years ago. A poll by NOP found that 48 per cent will eat GM food and 44 per cent still refuse. Only 20 per cent believe it is significantly less safe. Last year 50 per cent rejected GM food while 46 per cent ate it.

As the DETR (April 3, 2001) announced that new trial sites for genetically modified (GM) maize (a national survey conducted by NOP) reveals that support for GM foods amongst the British public has increased over the last 12 months. Key findings include:

1. Half the nation would eat GM food A 10 per cent decrease in those who believe GM foods are unsafe.
2. Two out of three people do not feel they know enough about GM foods Contrary to claims that there is no demand for GM food in the UK, and major campaigns against it, 48 per cent of people questioned said that they would eat food they knew was genetically modified. 44 per cent of people questioned said they would not.
3. Comparisons with an identical poll in 2000 demonstrate an increase in consumer acceptance. Then, 46 per cent said they would and 50 per cent said they would not eat it. The greatest level of support resided in the north east of England, where 60 per cent of respondents expressed a willingness to eat GM products: broadly the divide is between willing young men in the north east and reluctant middle aged women in the west country and Wales. A significant swing in the dynamics of the GM debate was also apparent in the results of the survey. While in 2000, 30 per cent of people questioned thought GM foods were not as safe as conventional foods, this year, it was only 20 per cent.

Professor Vivian Moses, Chairman of the CropGen panel that commissioned the survey comments: *This survey reveals that the changing attitudes of the public as the potential of GM crops are being recognised. It is clear that the public wish to make up their minds on this issue; however with 69 per cent of people saying they are not informed enough to make a judgment on the benefits and risks of GM crops, there is a clear demand for information. The farmscale trials offer an opportunity to gain such valuable information and will prove vital in evaluating this technology.*

The Frankenbuck Campaign against Starbucks

The Frankenbuck campaign is expanding internationally and is keeping up the pressure in the USA. On March 20, 2001, several thousand Frankenbuck volunteers took to the streets and sidewalks across North America, picketing, protesting, and leafleting in front of Starbucks coffee shops in over 100 cities. In a number of cities and college towns, activists staged press conferences enlivened by colorful signs, street theater, and powerful speakers. In Seattle 100 protesters rallied and cheered the speeches of campaign spokespersons--in front of a solid bank of TV cameras and news reporters--outside the building where Starbucks was holding their annual shareholders meeting. Since March 20, 2001 the Frankenbuck Campaign will continued in a number of communities and soon catch fire in Canada, Japan, and the UK.

Extensive Media Coverage against GMO

The activities of the Starbucks campaign have already generated over 750 newspaper articles, TV news clips, and radio reports against GMOs. Over 50,000 Consumer Warning leaflets have now been distributed to Starbucks' customers by our campaign. Millions of people across the US, Canada, and Europe have now heard about the campaign, and of course this negative publicity has Starbucks worried.

A national press release against this largest consumer campaign (March 20, 2001) ever launched in the US around the issues of genetic engineering and Fair Trade. Starbucks began to react in a manner quite seriously.

Panel OKS bill to crack down on eco-terrorists

TALLAHASSEE, Fla. (AP) - Activists opposed to bioengineered foods who destroy crops in protest could be sued for twice the value of the crop under a bill the Senate Judiciary Committee unanimously approved (Associated Press, April 17, 2001). The measure also would make it a third-degree felony carrying a possible prison term of three years to trespass on lands used for testing and researching food biotechnology. In the last three years, various groups have damaged or destroyed more than 40 private and government properties across the country where genetic engineering was being conducted, according to an analysis by Senate staff. The legislation must still clear the Appropriations Committee before making it to the floor.

12

Transgenic Plants: Promise and Danger

With the current trends of population growth and agricultural production, the demand for food in the most populous parts of the World will double by the 2025, and nearly triple by 2050. Increases in food productivity of this magnitude can not be brought about in a such a short period of time by conventional breeding, especially when some of our most important crops are approaching the physiological limits of productivity, nor increasing the amount of arable land, which accounts for 97 per cent of all food production in the world. The only solution to this problem is the development of transgenic crops. It is clear that during the next few decades a wide variety of transgenic crops will become integrated into agricultural systems in the industrialized countries. Introduction to such crops into the developing countries, which need them most, will be slow and largely through the efforts of multinational biotechnology companies, because most of the developing countries currently lack the scientific and industrial infrastructure to develop and introduce these technologies into their agriculture (Vasil, 1998).

From a genetic perspective, concerns that biotechnology is "unnatural" ignore that dynamic changes in the genetic code that occur within and across species genomes through modification of transposable genes or elements by virus vectors and through mutation. This article has attemptted to point out that humans have already had a huge effect on the genetic structure of many plants and animals, using many different technologies which will continue to develop and change overtime. However, while those in the forefront of any technology will promote its potential benefits, in the end it will be the economic and regulatory systems of governing bodies at the national and global levels that must evaluate the technology's relevance and appropriateness.

Advance in recombinant DNA techniques have extended the possibilities for the production and distribution of transgenic crops. Genetic engineering techniques have moved from the research laboratory to the applied sector. As a result, genetically engineered microorganisms are being used in large-scale industrial settings to produce pharmaceutical and other products. In agriculture, successful research and development has produced transgenic plants, animals, and microorganisms intended for use in the environment. Other potential environmental uses of genetically engineered microorganisms include cleaning up toxic wastes, mining copper and uranium, and improving oil recovery processes.

The recombinant organisms developed for environmental use inevitably have to be tested in the field, However, the prospect of extensive field tests of genetically engineered organisms has given rise to concern about possible risks these activities may pose to humans and the environment. This concern, which at times has resulted in acrimonious debate, has been shared by scientists as well as the general public.

One of the foundations of the scientific debate is the quite different viewpoints of scientists trained in different scientific specialties, such as molecular biology and ecology. Another cause of the scientific debate is the lack of appropriate data and validated conceptual or mathematical models for the processes under examination. In addition, there is not yet a commonly agreed upon overarching framework with which to integrate and evaluate the data and models that do exist.

Concern about risks to human health and the environment is not peculiar to biotechnology. Rather, these questions have emerged as an important component in the development, regulation, and production of the products of many new and older technologies such as chemicals and pharmaceuticals. Biotechnology is a part of this trend. The new scientific discipline of risk assessment has emerged to better assess the likelihood of hazards and risks associated with materials in the environment. In this volume, we have applied some of the tools and concepts of risk assessment to the evaluation of the risks associated with the environmental release of genetically engineered microorganisms and plants.

Scientific/Safety Issues

Has FDA's consultation process achieved its intended purpose? Based on experience to date, should this regulatory approach "sunset," continue in its current state, be made mandatory, or otherwise be revised?

FDA's so-called *consultation process* has failed to assure public safety and it has helped to undermine confidence in the products of U.S. agriculture. The process should be scrapped immediately and reformulated in the context of a new, unified regulatory system.

What newly emerging scientific information related to the safety of foods derived from bioengineered plants is there, if any?

Are there specific tests which, if conducted on such foods, would provide increased assurance of safety for man or animals consuming these foods?

There is substantial new information related to the safety of bioengineered foods, as well as old information that the agency has apparently ignored. New information includes emerging knowledge of the massive and ubiquitous effects from transgenic plants in the agricultural environment and livestock physiology. These effects may indeed have important implications for food safety and they have not been recognized by any U.S. regulatory agency. Mandatory pre-market safety testing must be based on the actual transgenic product, as used in the real world and as consumed by real people and animals. In addition to the requirement for real-world pre-market testing, there must be much more rigorous controls on the transparency and objectivity of the testing that is done.

What types of food products derived from bioengineered plants are planned for the future?

Will these foods raise food safety issues that would require different approaches to safety testing and agency oversight? If so, what are those approaches?

The powerful and uncertain qualities of current and planned transgenic food *events* demand a complete overhaul of the governmental approach to safety testing and agency oversight, as we have stated.

Public Information Issues

Should FDA's policy requiring labeling for significant changes, including changes in nutrients or the introduction of allergens, be maintained or modified?

Should FDA maintain or revise its policy that the name of the new food be changed when the common or usual name for the traditional counterpart no longer applies? Have these policies regarding the labeling of these foods served the public?

FDA's labeling policy has failed to serve the public interest. The failure to require labeling is imposing unacceptable costs and risks on the farmers and manufacturers who do not wish to have anything to do with transgenic foods. We are bearing the demand of consumers to know what is in their food. The public has a right to require labeling of transgenic content in their foods, and the government must implement this in a way that unequivocally places the costs and liabilities on the patent-holders and their licensees.

Should additional information be made available to the public about foods derived from bioengineered plants? If so, what information?

Who should be responsible for communicating such information?

All data and background information related to safety, efficacy, use and fate in the environment of transgenic food products and ingredients must be provided by the product developers and made available to the public. In addition, FDA must go out of its way to document and explain all information about the regulatory actions of the agency, including conflict-of-interest recusals by the numerous FDA personnel who have worked in the biotechnology sector.

How should additional information be made available to the public: e.g., on the Internet, through food information phone lines, on food labels, or by other means?

FDA should use every means at its disposal to distribute information to the public, and it must also make a much stronger effort to include the public in its decision-making. Above all it must heed what it is hearing from consumers and contribute to overhauling the entire regulatory system for *bioengineered foods.*

GROWTH OF BIOTECHNOLOGY PRODUCTS REQUIRE REGULATION

The increase in the number of biotechnology products subject to regulation requires growth in staff and capabilities in the relevant regulatory agencies. Several measures, including the number of patent applications, have been used to predict the increase in personnel that will be necessary to ensure the timely processing of permit applications and premanfacturing notices. Sufficient data are already available to demonstrate that growth in permit applications has been rapid, indeed exponential, throughout the 1980s, and to point out where problems lie.

Table 12.1 is a summary of a survey of biotechnology patents issued between 1980 and 1984 that could result in permit applications or premanufacturing notices to the USEPA.

Table 12.1: Microbial Patents Issued, 1980-1984

Area	No. of genera	No. of patents
Agricultural chemistry	8	37
Conversion of biomass	14	809
Industrial chemical production	14	528
Monitoring measurement, biosensor	7	36
Energy	4	217
Polymer or macromolecule production	16	787
Enhanced oil recovery	9	25
Waste and/or pollutant degradation	17	541
Mining and metals recovery	10	80

More than 3000 patents were issued during the 5-years period, and the paten office had a backlog of 7000 patent applications that have not been included in the survey. Thus, a total of 10,000 patents were applied for or granted between 1980 and 1984. The impact of these patents on the USEPA and federal agencies is just beginning to be evident.

Table 12.2: Coordinated Framework-Approval of Commercial Biotech. Products

Subject	Responsible agency (or agencies)
Foods and food additives	FDA[a], FSIS[b]
Human drugs, medical devices, and biologics	FDA
Animal drugs	FDA
Animal biologics	APHIS
Other contained uses	USEPA
Plants and animals	APHIS[a], FSIS[b], FDA[c]
Pesticide microorganisms released in the environment	USEPA[a], APHIS[d]
Other uses (microorganisms)	USEPA[a], APHIS[d]
Intergeneric combination	
Intrageneric combination	
Pathogenic source organism:	
Agricultural use	APHIS
Nonagricultural use	USEPA[a,e], APHIS[d]
No pathogenic source organisms	USEPA report
Nonengineered pathogens	
Agricultural use	APHIS
Nonagricultural use	USEPA[a], APHIS[d]
Nonengineered nonpathogens	USEPA report

Source: OSTP 1986

[a] Lead agency.
[b] FSIS, Food Safety and Inspection Services, under the Assistant Secretary of Agriculture for Marketing and Inspection Services, is responsible for food use.
Marketing and Inspection Services, is responsible for food use.
[c] FDA is involved when in relation to a food use.
[d] APHIS, Animal and Plant Health Inspection Service, is involved when the microorganism is plant pest, animal pathogen, or regulated article requiring a permit.
[e] USEPA requirements will only apply to environment release under a "significant new use rule" that EPA intends to propose.

Table 12.3: Summary of Releases of Engineered Organisms

	No. of genera	No. of tests	Major ecological issues
Plants	6	52	Gene transfer presence of related plants, exposure of endangered species, possibility of enhanced herbicide use, insect resistance to pesticide, weed resistance to herbicides, use of antibiotic resistance genes as markers
Microbes	14	56	Gene transfer and impact on microbial community structure, exposure of endangered species, alteration of host range, competitive ability of recombinant, use of antibiotic resistance genes as markers, mechanism of dispersal (or transmission) of microbe

POTENTIAL HAZARDS

The potential hazards are of different types. There would be reason for alarm if the recombinant molecule becomes dangerous in some way. The new properties it acquires may have side effects. The inserted gene fragment might possibly code for a harmful (or) toxic proteins. It might cause disease (or) it might transfer genetic material to human cells and made them cancerous in the **shot-gun experiment**, the entire DNA of an organism is fragmeted and inserted, although only one gene is desired. It is possible that an unwanted segment might cause harm. Suppose that genes specifying resistance to various antibiotics were put in to *E. coli* if by accident the new combinations of drug resistance would be transferred to pathogenic microbes. It would create some new disease for man. It was due to this concern, scientists became restricted on their research in the human health and society. This self-imposed restriction are known as the **moratorium.**

Transgenic Plants with New Agronomic Traits

Two different techniques are commonly used to introduce genetic information (DNA) inside cells, protected by pectocellulosic walls. A particle bombardment process, called biolistic, is used with success for monocots gynmosperm, squash, peas,...Today, this technique allows the bombardment of meristems and avoids the difficulties of regeneration. But the mediation of Agrobacterium remains till now the most routinely system to transform dicotyledonous plants. The main agronomic traits introduced in horticultural plants and already commercialised are Bt toxin and herbicide resistance. Other studies concern virus resistance, male sterility.

Transmission of the New Traits

These traits induced by the transferred DNA are transmitted to the progeny as a dominant Mendelian character. Nevertheless in some cases, transgenic plants with a strong gene expression can generate progenies with only a faint no gene expression. This problem is not clearly elucidate.

Marker, Reporter, Promotor and Expression Genes

To select the transformed cells, some marker genes, very often resistant to antibiotics or to herbicides, are attached to the coding sequence, as a promoter or an expression gene. These allow the new gene to express in the whole plant or in a specific plant tissue.

Some reporter genes are used to follow the evolution of the transformed cells. The most frequent reporter genes are the gene "gus" (- glucoronidase) or the gene lux (luciferase).

Horticultural Transgenic Crops Already Released in USA

The Flavr Savr tomato was the first genetically engineered whole food approved for commercial sale in 1994. Four other transgenic tomatoes with a delayed ripening were approved later (1995 and 1996).

In 1995, potatoes with Bt genes and a squash cultivar resistant to two viruses were released. Two years later, another squash cultivar resistant to 3 viruses and a papaya line also resistant to viruses were approved.

Two other horticultural crops are waiting the authorities approval: red hearted cichory (Radicchio) with male sterility and resistance to herbicide, and a tomato with Bt toxin.

It there are so few commercial transgenic plants in horticulture, it's probably due to the relatively low acreage of these horticultural crops in comparison with other agricultural crops. In 1997, in the world 5 million ha were planted with glyphosate resistant soybeans released by Monsanto alone.

Nowadays, it's true also that many transgenic plants exist in research laboratories, awaiting authorization for field testing, as they are very interesting model plants for learning physiology: lettuce with less nitrate by increasing nitrate reductase gene activity, a yeast ribonuclease gene to reduce viroid infection on potato, modification of lignin synthesis in plant to produce trees adapted to paper industry, or timber, or biomass production, regulation genes to control tree architecture.

THE EFFORTS

The debate over the *safety* of using genetically engineered organisms in the environment has been marked by public controversy and la large and a large degree of polarization. The initial focus of the debate was agricultural uses of plants and microorganisms; however, the debate was really of importance to all end uses of environmental biotechnology, including bioremediation, oil recovery, coal desulfurization, mining, and others.

In the scientific literature, the safety debate was highly visible in the pages of *Science* magazine. Winston Brill, a microbiologist with both an academic and industrial biotechnology background, published an article on safety concerns and genetic engineering in agriculture (Brill 1985) in which he argued that plants and animals have been altered by traditional breeding by humans for centuries without any serious problems. In addition, bacteria and fungi, including pathogenic organisms, have been added to soils and plans in an effort to find beneficial uses of these organisms. Brill concluded that these observations compared an adequate and appropriate data base from which to draw risk assessment conclusions, including for genetically engineered organisms where only a few genes have been added.

Brill's article generated a response from ecologists Robert Colwell, Elliott, Norse , David Pimentel, France Sharples, and Daniel Simberloff (Colwell et al. 1985), who presented a very different view of the risks. Colwell et al. took Brill to task for his qualitative risk views of *seems very small* and *is extremely unlikely* and suggested that quantitative risk assessment is both necessary and appropriate. The ecologists argued that mutations intended to increase the range of conditions under which an organism is intended to survive are not small from an ecological point of view. They also argued that the phenotype of a microorganism, especially its ecological traits and population dynamics, is not fully predictable from genotype alone. Finally, Colwell et al. (1985) advocated assessing the risk of releasing genetically engineered organisms into the environment on a case-by-case basis.

In a policy forum set up by *Science* magazine a few years later, side-by-side articles by ecologist Frances Sharples (1987) and bacteriologist and molecular biologist Bernard Davis (1987) were published.

Sharples (1987) continued the ecologists' argument that risks associated with the environmental release of engineered organisms must be evaluated on a case-by-case basis. She noted that while the "absence of human health effects of biotechnology workers in contained environments is admirable, it is not particularly relevant to the question of whether uncontained uses of modified organisms will be equally harmless." Sharples pointed out that "shifts in environmental contexts may be as important as genetic modifications in determining whether the ecological relationships of an engineered organisms will be unique relative to those of a parental form."

Davis (1987) reiterated the molecular biologists' arguments that the use of modified microbes is not really novel, "but is an extension of the old process of domestication of wild organisms-including the selection of microbial variants to make bread, wine, antibiotics, or vaccines." Davis also argued that the *experience of ecologists with transplanted higher organisms is less pertinent than are the insights of fields closer to the specific properties of engineered microorganisms: population genetics, bacterial physiology, epidemiology, and the study of pathogenesis.*

Discussions and debates between molecular biologists and ecologists have also proceeded via symposia by professional societies. For example, in 1985, the American Society for Microbiology sponsored a cross-disciplinary symposium entitled Engineered Organisms in the Environment: Scientific Issues. The published proceedings of that conference (Halvorson et al. 1985) remain a frequently cited reference regarding scientific considerations and uncertainties in risk assessment. An international conference on the Release of Genetically Engineered Microorganisms (REGEM I) was held in South Wales in 1988. The conference proceedings (Sussman et al. 1988) provide many view points regarding what is necessary to ensure the safety of environmental release experiments.

The U.S. National Academy of Sciences (NAS) published a pamphlet intended for a general scientific audience in 1987 entitled *Introduction of Recombinant DNA-Engineered Organisms into the Environment: Key Issues* (NAS 1987). The NAS made some general conclusions regarding risk, including:

> *There is no evidence that unique hazards exist either from the use of recombinant DNA techniques or from the movement of genes between unrelated organisms.*
>
> *The risks associated with introduction of genetically engineered organisms carrying recombinant DNA are the same in kind as those associated with the introduction of unmodified organisms and organisms modified by other methods.*
>
> *Assessment of the risks of introducing genetically engineered organisms carrying recombinant DNA into the environment should be based on the nature of the organism and the environment into which it is introduced, not on the method by which it was produced.*

While these points, especially the third one, were common in the mainstream discussion of potential risks and regulation, the pamphlet drew protests from scientists (and others) with environmental concerns because of the lack of documentation of the assertions in the report.

In 1989, the NAS published a second report on environmental release, entitles *Field Testing Genetically Modified Organisms: Framework for Decisions,* (NAS 1989).This report was more detailed than the 1987 pamphlet, and was intended for more technical audiences. In a view of risk assessment that differs quite a bit from that used for chemical risk assessment, the NAS posed three questions to use in making initial judgments of risk, and thus suggested the extent of the informational requirements for risk assessment (NAS 1989):

> *Are we familiar with the properties of the organism and the environment into which it may be introduced?*
>
> *Can we confine or control the organism effectively?*
>
> *What are the probable effects on the environment should the introduced organisms or a genetic trait persist longer than intended or spread nontarget environments?*

At approximately the same time as the 1989 NAS report appeared, the Ecological Society of America (ESA) published a special feature entitled "The Planned Introduction of Genetically Engineered Organisms. Ecological Considerations and Recommendations" (Tiedje et al. 1989). It focused on the ecological and evolutionary aspects of planned introductions of genetically engineered organisms, and included microorganisms, plants, and animals.

The ESA report set forth a preliminary set of specific criteria for the scaling of regulatory overnight. Attributes of organism sand environments were divided into four categories: (1) genetic alteration, (2) parent (wild type) organism, (3) phenotypic attributes of engineered organism in comparison with parent organism, and (4) environment. For each category of attribute, the authors defined numerous specific attributes, and a scale on which to base the level of scientific consideration necessary scrutiny be commensurate with the level suggested by the scientific attribute scales.

From the preceding discussion, it is clear that the scientific debate over the safety of the environmental release of genetically engineered organisms has evolved greatly since its initial stages. In the mid-1980s, people argued that their view of the world was absolutely correct, and cited historical examples that supported their points. In the late 1980s, more thoughtful discussions took place about what considerations are important for risk assessment, how to develop overaching risk assessment frameworks, and how to scale regulatory scrutiny on the basis of risk. It should be noted that throughout this period, governmental, academic, and other groups were examining the scientific need for risk assessment.

Risk assessment can play a key role in the regulation of environmental biotechnology. Work on scaling of scientific scrutiny can assist in the determination of what organisms should be regulated, and with what level of attention. The choice of overarching framework for risk assessment will have a profound impact on the type of data required for the risk evaluation, and thus what data will be required for permit applications. However, these risk assessment discussions may not solve the problems of public acceptance of biotechnology; this is an issue that must be handled within the larger social and economic context.

In putting together this volume, it has been our intention to extend the discussion of what we know and what we don't know about how to predict the safety of environmental releases of genetically engineered plants and microorganisms. We hope that this book will stimulate further advances in thinking about risk assessment as it relates to this important field. Although most of the chapters focus on purely scientific or regulatory needs in risk assessment, we have also included two chapters on the social and international aspects of environmental biotechnology. These topics are likely to be increasingly interwoven into discussions of the risks associated with the commercialization of environmental biotechnology.

NO CONCLUSIONS YET ON ECO EFFECTS OF GM ORGANISMS

More than a decade after the US Government allowed the first release of a genetically engineered organism into the environment, researchers have now concluded that scientists still cannot say with any precision what the ecological effects – either good or bad – of such genetically modified organisms might be. The findings, published in *Science,* raise questions about why so little is known and whether some key questions about risk are, in practical terms, answerable (Business Line, Dec. 30, 2000)

For example, some scientists have estimated that answering just a single question of risk for a single organism – whether a type of bio-tech corn harms the monarch butterfuly – would cost $2-3 million, more than Agriculture Department typically grants each year for the study of environmental risk. And if questions cannot be answered, where do we go from here?

Questions about risk first emerged when genetically engineered organisms started making their way out of the laboratory and into the public consciousness in the late 1980s.

Men in spacesuits were assigned to release the first genetically engineered organisms into the wide world: bacteria sprayed on strawberries to protect them from frost. Soon afterward uneasy shoppers shunned the first genetically engineered crop, the ill-fated Flavr Savr tomato, whose only crime was a foreign gene for longer shelf-life.

Today such organisms seem almost quaint as biotech salmon grow to market size in half the normal time and genetically modified goats make human blood proteins in their milk. And an international debate has sprung up over the value to these organisms, with participation from such unlikely quarters as the Vatican and Prince Charles.

Biotechnology has raced ahead, scientists, ability to predict potential environmental consequences apparently has not, according to the new *Science* paper, a review of scientific literature by Dr. LaReesa Wolfenbarger and Dr. Paul Phifer. The study of the highest profile of environmental risks, the potential threat of genetically modified corn to monarch butterflies, is a case in point.

Questions about corn and monarchs first arose in the spring of 1999 when Cornell researchers showed that monarch caterpillars died in the laboratory after eating pollen from genetically engineered corn. The corn, given a gene from the *Bacillus thuringiensis* bacterium, then produced a toxin that killed the European corn borer pest.

Corn and monarch butterflies are two of the best studied organisms on the planet. How difficult could it be to determine whether monarchs were indeed at risk in the wild? But last month, after a year-and-a-half of research by more than 20 researchers from universities and industry, scientists gathered outside Chicago were still unable to say with any precision what the magnitude of risk was from the biotech corn to wild monarch populations.

One reason so little is known about the magnitude of ecological risks in general is that regulators deemed some effects, including those to species such as the monarch – which are neither beneficial to agriculture nor legally protected as endangered species – of little concern.

We knew things such as monarchs and other butterflies would be susceptible, said Dr. Arnold Foudin, an assistant director of scientific services at the Department of Agriculture. But the major problem for researchers is the inherent difficulty, expense and time involved in understanding ecological interactions.

Biologists first set out to see whether monarch caterpillars would even encounter Bt corn pollen in the wild. By fall, biologists announced that most toxic pollen is shed within a cornfield, rather than outside it. If monarchs were unlikely to live cornfields, as industry spokesmen suggested, the impacts of the new crop might be quite limited. But this summer researchers discovered that large numbers of monarch caterpillars live on milkweed in cornfields.

We're finding them between the rows of corn, said Dr Karen S. Oberhauser, ecologist at the University of Minnesota. To complicate matters further, risks to monarchs seem to vary geographically depending on when pollen is produced and monarchs are in abundance. Also, some Bt corn varieties are more toxic than others.

More important, other scientists report finding this summers that the survival of monarch caterpillars in Bt and normal corn fields is indistinguishable. The potential complication is that the vast majority of caterpillars are killed by predators. *Only two to five per cent of them ever make it*, said Dr Rick Hellmich, research entomologist with the Department of Agriculture working at Iowa State University.

So while some are again ready to conclude that Bt corn poses no undue risk to the monarch, others say that with such a minuscule survival rate, even important differences between the survival rates in the two fields would be tiny, probably requiring much larger-scale studies to detect. *Given the ups and down on this one, I'd hate to jump to conclusions now*, said Dr David Andow, entomologist at eh University of Minnesota.

The difficulty in coming to conclusions about corn and monarchs raises the question of how much scientists can be expected to learn about what researchers say are the other much more complex ecological threats from biotech organisms.

In addition, authors of the new *Science* paper say there are some risks that scientists may never be able to fully understand. For example, in the past scientists have recorded delays of 30 to 50 years between the arrival of a plant and its widespread infestation as a weed, making reliable predictions of the long-term likelihood of threats like superweeds extremely difficult.

Dr John Losey, a co-author of the original monarch study, said he and a group of other researchers originally estimated it would cost between $2 million and $3 million to answer the monarch question.

If this is the cost of understanding just one risk from one bio-tech organism to one species, where will the big research bucks come from to answer the range of questions being raised by the many new organisms being developed?

EFFORTS AT INTERNATIONAL LEVEL

The U. S. Regulatory System

The first federal coverage of the safety of biotechnology, perhaps one could stretch and call it "and this *is* stretch of the word since compliance was voluntary, grew out of the Asilomar Conferences and other meetings organized by scientists concerned about the potential biohazards associated with the use of the newly developing recombinant DNA techniques and the possible escape of the newly created microorganisms from the laboratory. The second Asilomar conference in 1975 was immediately followed by the establishment of the National Institutes of Health Recombinant DNA Advisory Committee (NIH RAC) and the development and publication of the well known RAC guidelines in 1976. It is interesting to note that these guidelines amounted to a regulatory scheme, since NIH-funded researchers had to comply for fear of jeopardizing major funding to their institutions. As other U.S. government funding agencies adopted the guidelines for their grantees, all federally funded molecular biological research was covered. Nonfederally funded researchers were expected to, and did, comply on a voluntary basis. Eventually, guidelines similar to those of the NIH RAC were applied to researchers around the world.

One outcome of this procedure was that the Institutional Biosafety Committees (IBCs) were established to aid the RAC by reviewing some applications at the local level. The IBCs became, and still remain, the reviewers of most of the molecular biology experiments, now called *biotechnology* experiments, requiring review in the United States, Needless to say, the RAC guidelines focused on laboratory worker safety. They functioned with great success via a commitment to containment of the engineered microorganism within the confines of the laboratory. In fact, in the initial version of the RAC guidelines, any type of environmental release was completely prohibited.

With the application of laboratory-based molecular biology research to industrial processes and products, questions regarding the safety of the technology once again emerged. This time, the health and safety of industrial workers, the environment, the general human population, and agricultural products all came under scrutiny. As commercialization began, federal agencies with the relevant regulatory authority became involved with biotechnology processes and products.

The overall policy regarding how biotechnology will be regulated and coordinated in the United States has been set forth in formal notices published in the *Federal Register* (OSTP 1984, 1986). Three agencies share primary responsibility for regulating the organisms and products of recombinant DNA biotechnologies, whether they be designed for closed systems or for environmental uses. They are the USEPA, the FDA, and the USDA. In addition, the Occupational Safety and Health Administration (OSHA) has the responsibility for ensuring the health and safety of biotechnology workers, and the Department of Health and Human Services for ensuring the health of the general public.

In addition to federal regulation, approximately 13 states and several municipalities have enacted biotechnology-related legislation, including provisions related to the environmental release of genetically modified microorganisms. The most comprehensive of these laws was enacted by North Carolina in 1989. State and local laws have been reviewed and analyzed elsewhere (Industrial Biotechnology Association 1990, Fox 1989, Krimsky et al. 1982).

Each of the federal agencies regulating biotechnology is guided in its analysis and decision-making criteria by its specific legislation-i.e., the laws passed by Congress charging each agency with specific responsibilities. These laws differ in their mandate as to what populations to consider with regard to adverse effects (e.g., humans, crops, the environment), as well as in their mandate as to how to strike a balance between risks and benefits. In addition to its specific legislation, each agency must also adhere to the National Environmental Policy Act (NEPA), which is binding on all federal agencies.

All of the legislation used to regulate recombinant DNA includes safety as an important element in decision making. However, none of the laws used to regulate biotechnology products specifically address the safety of biotechnology. Instead, the laws are based on the uses of the products, or the environmental discharge of materials. For example, the environmental protection laws administered by USEPA were written with inorganic and organic chemicals in mind. The laws cover the uses of products (pesticides, chemical inter-mediates, and many others), the emission of materials into the air and water, and the disposal of solid and hazardous waste. At the USDA, the laws were written with naturally occurring organisms, and especially plant and animal pathogens or toxins, in mind.

The goal of NEPA is to assure that federal actions are environmentally sound. In essence, NEPA requires all federal agencies to evaluate the possible environmental outcomes of their proposed actions and to look for a balance between benefits and possible adverse impacts. Agencies must conduct a thorough benefits and possible adverse information and alternatives and seek public comment. At one time, agencies were required to consider worst case scenarios in their NEPA reviews of biotechnology, but this is no longer case.

U. S. Environmental Protection Agency

The USEPA administers seven environmental statutes. It specifically regulates biotechnology under two of them: the Toxic Substance Control Act (TSCA), administered by the Office of Toxic Substance (OTS), and the Federal Insecticide, Fungicide, and Rodenticide Act (FIFRA), administered by the Office of Pesticide Programs (OPP). These two statues have been described as *gateway* legislation. They are not abatement-oriented as are other statues administered by the USEPA (e.g., the Comprehensive Environmental Response, Compensation, and Liability Act, also known as CERCLA or the Superfund law), but instead they are preventative. They are invoked before application of a new product to the environment. FIFRA covers all pesticidal products and is clearly applicable to biotechnology products. TSCA is invoked through the concept of "newness." New chemicals or new uses for existing chemicals trigger a TSCA review. When asked whether TSCA applied to microbial products or recombinant DNA, the USEPA's General Counsel wrote in 1983 that since "TSCA covers all chemicals and DNA is a chemical, DNA is covered" and added that "a DNA molecule does not have any use except in the life form of which it is part" (U.S. Congress 1984). This position has not been challenged. USEPA reviews under TSCA and FIFRA have been accepted in court as the equivalent of NEPA reviews.

The USEPA regulates wastes resulting from the biotechnology industry under its waster-related laws, such as the Resource Conservation and Recovery Act (RCRA) for hazardous and industrial solid wastes, and the Clean Water Act for wastewater discharges. No special provisions have been made for biotechnology. Regulations and policies affecting the developing of genetically engineered microorganisms intended to clean up the environment (bioremediation) at toxic waste sites have been made under CERCLA, but these have been directed at new technologies in general, rather than biotechnology in particular.

In addition to regulating biotechnology, the USEPA in 1983 formally initiated a Biotechnology Risk Assessment research program in its Office of Research and Development to develop methodology for biotechnology risk assessment. The program was intended to develop methods required for regulatory decision making. It covers the engineered microbes, the DNA involved and the potential for environmental and human health effects. The research includes work on modifying existing or developing new methods to identify, monitor, and track engineered microbes and recombinant DNA in field situations, as well as work geared to developing methods to identify and quantify environmental and health effects (Levin et al. 1987).

Between 1983 and 1989, the OPP and OTS offices of the USEPA reviewed or were in the process of reviewing approximately 56 biotechnology products. As can be seen, 55% of the reviews involved just two bacterial species, *Rhizobium meliloti* and *Bacillus thuringiensis,* either as the host or donor. *R. meliloti,* has been reviewed by OTS under TSCA authority as a new chemical, and *B. thurngiensis* by OPP under FIFRA authority as a pesticide. The USEPA's Biotechnology Science Advisory Committee reviews the actions of both OPP and OTS and attempts to coordinate and unify the procedures of both offices.

U. S. Department of Agriculture

The USDA has responsibility for food and fiber products. It is in one sense outside the circle of ecological safety, in that its primary concerns are the safety of crop plants and food animals, and the safety and wholesomeness of food products.

Table 12.4: Recombinant Microbe Tests

Microbe (no. of tests)	Trait (no. of tests)	Source of genetic material
Rhizobium meliloti (20)	Nitrogen fixation (11) Marker gene (8)	Bacteria
Bacillus thuringiensis (11)	Insecticide ability (11)	Other *B. thuringiensis*
Bradyrhizobium japonicum (6)	Marker gene (4) Nitrogen fixation (2)	Bacteria
Pseudomonas fluorescens (4)	Frost protection (2) Insect resistance (2)	Gene deletion, *B. thuringiensis*
Trichoderma harzianum (4)	Host range (4)	Other trichoderma, undirected mutagenesis
Clavibacter xyli (2)	Insects resistance (2)	*B. thuringiensis*
P. syringae (2)	Frost protection (1) Insecticide ability (1)	Gene deletion, other *P. syringae*
Colletotrichum gloeosporiedes (2)	Host range (2)	None: chemical mutagenesis
P. aureofaciens (1)	Marker gene (1)	*Escherichia coli*
Vaccinia virus (1)	Immunization potential (1)	Virus (rabies)
Sclerotinia sclerotivorum (1)	Host range (1)	None: mutagenesis
T. viride (1)	Host range (1)	None: undirected mutagenesis
Autographa colifornica (1)	Persistence (1)	Gene deletion

The USDA has three arms that deal with biotechnology: the Agricultural Research Services (ARS), the Food Safety and Inspection Service (FSIS), and the Animal and Plant Health Inspection Service (APHIS). The ARS focuses on research issues and has formed a group equivalent to the NIH RAC, called the Agricultural Biotechnology Recombinant DNA Advisory Committee (ABRAC), to review proposals. The ARS is also instituting an information service as part of its National Biological Impact Assessment Program. The FSIS functions to assure the safety and the wholesome characteristics of food products. Via APHIS, the USDA meets its responsibilities for licensing veterinary biological material and issuing permits for transport of biological material. APHIS has formed the Biotechnology, Biologics, and Environmental Protection Division (BBEPD) with responsibility for all biotechnology products.

Table 12.5: Recombinant Plant Tests.

Plant (no. of tests)	Trait (no. of tests)	Source of genetic material
Tomato (22)	Herbicide resistance (8)	Bacteria, virus, tobacco, petunia, soybean, pea
	Insect resistance (8)	*B. thuringiensis*
	Ripening delay (3)	Bacteria, tomato, virus
	Resistance to mosaic virus (3)	Virus, bacteria, pea
Tobacco (12)	Herbicide resistance (4)	Bacteria, virus, tobacco
	Would enzyme (2)	Potato, bacteria
	Insect resistance (2)	Cowpea, virus, *B. thuringiensis*
	Resistance to mosaic virus (1)	Virus, bacteria
	Resistance to fungi (1)	Bacteria, bean
	Reduced heavy metals (1)	Mouse, bacteria, virus, pea
	Increased lysine synthesis (1)	Bacteria, virus, pea
Cotton (6)	Herbicide resistance (1)	Petunia, bacteria, virus, soybean, tobacco
	Insect resistance (1)	B. thuringiensis
Soybean (3)	Herbicide resistance (3)	Petunia, virus, bacteria
Potato (2)	Resistance to potato virus (2)	Virus, bacteria
Alfalfa (3)	Herbicide resistance (2)	Bacteria, virus
	Resistance to mosaic virus (1)	Bacteria, virus
Walnut (1)	Marker gene (1)	Bacteria
Poplar (1)	Wound enzyme (1)	Bacteria, potato
Cucumber(1)	Resistance to mosaic virus (1)	Bacteria, virus

Under NEPA, the USDA has a responsibility for ensuring safe ecological utilization of crops, livestock and veterinary products, whether they are produced by genetic engineering or traditional means.

Table 12.5 (above) provides a summary of the 56 field tests of recombinant plants that entered into the USDA process between 1983 and 1989. The USEPA and USDA have collaborated on the review of 23 of the engineered plants, mostly tobacco and tomato. The table shows that the source of genetic material is highly varied, and includes other plant genera, bacteria, viruses, and mammals. Field releases of transgenic plants have also been reviewed and approved in several countries outside the United States. For example, the Canadian government authorized 39 field trials of engineered plants during this time period, mostly with canola and flax.

FOOD AND DRUG ADMINISTRATION

The FDA regulates biotechnology under the authority of the Food, Drug, and Cosmetic Act (FDCA) and the Public Health Services Act. The agency has a mandate to ensure efficacy and safety of food and pharmaceutical products which are segregated into a number of classes. It has estimated that the FDA has already ruled on thousands of biotechnology products, although not all agree on what is required to label a product **biotechnology**. FDA's criteria for product evaluation, as set forth in the FDCA, revolve around purity, lack of adverse effects, and efficacy. FDA's population of concern regarding adverse effects is humans. FDA is also responsible for administering the tolerance limits for pesticides used on crops that are set by other agencies (e.g., USEPA).

The FDA has a major responsibility in biotechnology in that 60% of the current market share of biotechnology products passes through the agency for review. However, the FDA's environmental concerns and responsibilities are only in regard to NEPA. The USEPA is responsible for ecological and related public health issues.

OCCUPATIONAL SAFETY AND HEALTH ADMINISTRATION

OSHA has broad responsibility for worker safety. However, OSHA's active involvement in biotechnology regulation is minor. Existing guidelines and practices are seen as adequate to the need.

A committee to monitor developments in biotechnology that could affect worker health and safety has been established within the National Institute for Occupational Safety and Health (NIOSH). However, NIOSH serves OSHA only in an advisory capacity. Administratively, NIOSH and OSHA are located in separate departments (NIOSH is in the Public Health Service in the Department of Health and Human Services, while OSHA is in the Department of Labor).

COORDINATED FEDERAL FRAMEWORKS

There are two important dates to remember in connection with coordination of biotechnology regulation among U. S. agencies: 1984, the Office of Science and Technology Policy (OSTP) published a proposed coordinated framework for biotechnology that resulted in the formation of the Biotechnology Science Coordinating Committee (BSCC) (OSTA 1984). In 1986, the BSCC published in the *Federal Register* many elements of an altered proposed framework, a matrix of applicable legislation, and responsibilities for overnight of biotechnology products (OSTP 1986). The revisions were based, in part, on comments received on the 1984 proposal.

Table 12.3, adopted from the 1986 *Federal Register* notice, summarizes the scheme of regulatory authority that is currently operative in the United States. The table indicates the lead and cooperating agencies for the various uses of biotechnology products. Note that there is overlap between agencies in the foods and food additives area, plants and animals area, and in pesticides. In each of these areas of overlap, a lead agency has been designated. A case-by-case approach has been used to determine responsibility of overnight of other uses of microorganisms. The BSCC has been involved in resolving differences in definitions and in establishing communication between agencies, as well as dealing with international issues.

INDIA'S EFFORTS

Genetic engineering has the potential to revolutionise Indian agriculture but there is an urgent need of public debate on terminator genes, genetically modified corps and *Frankenstein foods*, etc... People do not trust on their Governments on their release, whether they are safe or not? India, like other countries, stands to benefit enormously from harnessing the power of genetic engineering and other developments in biology for agriculture, animal husbandry and treatment of disease. But national interest also dictates that the country has a sound and credible regulatory system to keep a close watch over the application of modern biotechnology and prevent its misuse.

Government of India already has regulations governing import, research, field trials and ultimate commercial use of genetically modified organisms or GMOs (organisms which carry an artificially introduced gene). There is a need of transparency in the functioning of the committees which perform these roles. India needs mechanisms to monitor biological patents being applied for and granted abroad as well as in India. This should not be left wholly to the Government and Indian NGOs need to keep an independent watch. The Indian Patent Office needs considerable improvement so that it can play its role of often being the first line of defence.

The scientific community and NGOs should became aware of the Terminator gene issue. North American NGO, blew the whistle on a U.S. patent innocuously titled **Control of Plant Gene Expression**. The technology provides a genetic method of making seeds sterile so that farmers can not save and replant them. Any paten granted in the U.S. and Europe provide warning signals each product, while India is dependent on foreign organisations.

The use of GMOs in India is governed by the Rules for the Manufacture, *Use/Import/Export and Storage of Hazardous Microoganisms/Genetically Engineered Organisms or Cells* promulgated in December 1989 by the Union Government's Ministry of Environment and Forests under the powers bestowed on it by the Indian Environment (protection) Act of 1986.

As the Terminator genes patent demonstrates vividly, the patent-title can be innocuous and even the abstract gives clue to its possible application. These rules established the following committees.

Committee	Governed by	Function
Recombinant DNA Advisory Committee (RDAC)	Department of Biotechnology, Government of India	To recommend appropriate safely regulations for genetic engineering work.
Review Committee on Genetic Manipulation (RCGM)	Department of Biotechnology Government of India	To monitor on going research. Every organisation handling genetically engineering organisms has to set up an institutional bio-safely committee
Genetic Engineering Approval Committee (GEAC)	Department of Environment, Forest and Wildlife, Government of India	to clear any large-scale use of genetically engineered organisms. These committees have been set up.
Technology Information, Forecasting and Assessment Council (TIFAC),	An autonomous body under the Department of Science and Technology	To notice an application of patent of terminator genes in India too.

Looking into the present situation and hazards of GMOs, there is a need to establish Biotechnology Coordination Committee in each state of India, headed by the Chief Secretary, with powers to inspect investigate and take punitive action in case of violations of statutory provisions. The rules also require district level committees, heated by the district Collector, to watch over issues of safety and meaning emergencies. In Karnataka, even the State Committee was established only after the Terminator gene controversy erupted.

According to Dr. Manju Sharma, Secretary to the Department of Biotechnology, the Review Committee on Genetic Manipulation (RCGM) only authorises research work on transgenics and genetically modified organisms. But research on transgenic plants requires laboratory as well as field experiments in contained glass houses and in small plots in the field. In plant experiments, without field trials in small plots it would not be possible to conclude how the surrounding flora and fauna would react to the transgenics. Once the green house experiments, limited experimental field trials and other experiments were completed, the scientific information from them would be complied. The REGM would then forward this, with its own observations, to the Genetic Engineering Approval Committee (GEAC) for a decision, said Dr. Sharma.

A Petition has already been field in the supreme Court, questioning the RCGM's authority to sanction field trials.

1. None of committees release information about the basis on which they have taken decisions about permitting or disallowing genetic engineering work. For example, when there are reports that Bollgard cotton has failed in other countries, information should be made available to scientists/general public and the reasons should be known to every body that why the tests in India were permitted. This question was raised by Dr. Pushpa Bhargava former Director of the Centre for Cellular and Molecular Biology.

2. There is no procedure for the committees to secure public comments. In contrast, the U. S. Department of Agriculture's Animal and Plant Health Inspection Service (APHIS) publishes a notice in the Federal Register (the U.S. Equivalent of the Indian Government's gazette) and seeks public review before it deregulates any genetically engineered plant or microorganism. When APHIS decides to go ahead and deregulate an genetically modified organism, its detailed determination of whether this organism could become a plant pest as well as an equally detailed environmental assessment of the potential impact of not regulating its use is made public. The APHIS Determination and Environment Assessment documents of Monsanto's Bollgard cotton are thus readily available.

In early 1999, M.S. Swaminathan Research Foundation hold a meeting on this issue and suggested that, in addition to the existing structure, there should also be a National Commission on Genetic Modifications of Crop Plants and Farm Animals, and should be headed by an independent and professionally qualified chairperson. He well coordinate and enforce a precautionary package for the safe and beneficial use of genetic engineering.

Much also needs to be done to improve the operations of the Indian Patents Office. A few years back, the patent Office actually cleared a patent which gave a U. S. company, Agracetus, rights over genetically engineered cotton when Indian law did not permit patenting of life forms. Once again, it was RAFI which drew attention to the patent! When the Indian patent law is modified to conform to the TRIPS requirements, it will need to allow genetic patenting. The role of the Patent Office will then become even more important.

According to Cornell's International service for the Acquisition of Agri-Biotech Application (ISAAA) pointed out that in China, with 1.8 million hectares land is under various genetically engineered crops, and is second only to the U.S. in this respect. Between 1996 and 1999, twelve countries, 8 industrial and 4 developing, have contributed to more than a twenty fold (23.5) increase in the global area of transgenic crops. Adoption rates for transgenic crops are unprecedented and are highest for any new technologies by agricultural industry standards. Higher adoption rates reflect grower satisfaction with the products that offer significant benefits ranging from more convenient and flexible crop management, higher productivity or net returns/hectare and a safer environment through decreased the of conventional pesticides, which collectively contribute to a more sustainable agriculture. In 1999, the global area of transgenic crops increased by 44%, or 12.1 million hectares, to 39.9 million hectares, from 27.8 million hectares in 1998. Seven transgenic crops were grown commercially in twelve countries in 1999, three of which, Portngal, Rumania and Ukraine, grew transgenic crops for the first time. According to ISAAA, the virus resistant tobacco introduced by China had 5-7 per cent greater yield than non-genetically modified tobacco and saved 2-3 applications of insecticide. When Bt cotton was grown on 1.8 million acres report. Bt cotton gave average yield increases of 7 per cent compared to non-be cotton and the yield increase could be as high a 20 per cent.

INTERNATIONAL BIOTECHNOLOGY GUIDELINES AND REGULATIONS

Many industrialized countries have instituted mechanisms for the regulation of biotechnology, including the environment release of genetically engineered organisms. Thus far, only a few of the more advanced developing countries have biosafety guidelines and regulations, and those appear to be only for contained uses. However, developing countries are also being increasingly faced with requests to conduct field trials of genetically engineered plants and microorganisms.

Several international organizations are developing, or have completed, guidelines related to biotechnology safety that have gained some measure of international consensus. These guidelines cover laboratory safety, contained industrial uses, and more recently, field trials of genetically engineered plants and microorganisms.

The Organization of Economic Cooperation and Development (OECD) has been the major forum thus far in the effort toward international harmonization of biotechnology regulations for both contained uses and environmental releases. The OECD's 1986 monograph, *Recombinant DNA*

Safety Considerations, although not binding on number countries, has had a major influence in the development of regulations in developed countries. A revision of these guidelines has been in the works for several years. In addition, a new set of OECD guidelines, "Good Developmental Practices," which covers protocols for field releases, was released in May 1990.

The European Community (EC) issued draft directives in the spring of 1988 on contained use of genetically modified microorganisms, and on the deliberate release to the environment of genetically modified microorganisms. These are still moving toward final form and passage in 1990. The final form of these directives will be binding on all EC countries, and will set minimum standards of regulation. However, each country will have to implement the directives individually, and may impose more stringent rules (Newmark 1988, Dixon 1989). Thus, environmental release regulations may differ substantially of microorganisms in EC countries range from the lack of any regulation in Italy and Spain to a nearly complete prohibition of environmental releases in the Federal Republic of Germany. In the United Kingdom, only notification of deliberate release experiments is required, although detailed guidelines are available to indicate how to balance the level of risk and the cost of taking precautions to avoid the risk. In Denmark, a pre-1989 near prohibition of environmental releases had ended, a field trial with genetically engineered sugar beets was approved in late 1989, and more permissive legislation may be enacted (Newmark 1988, Simpson 1990).

Several United Nations agencies, notably the World Health Organization (WHO), the United Nations Environment Program (UNEP), the United Nations Industrial Development Organization (UNIDO), and the Food and Agricultural Organization (FAO) have keen interests in biotechnology, including the biosafety and regulatory aspects of environmental biotechnology.

UNIDO, WHO, and UNEP have participated in a Working Group on Biotechnology Safety for several years. A document, *Safety Guidelines and Procedures for Bioscience-Based Industry and Other Applied Microbiology,* described the history and possible problems the Working Group would like to examine, and reviews the types of data available for assessing health risks, primarily to workers, in the biotechnology industry (UNIDO 1986). In 1989, the Working Group launched a program to develop strict new safety guidelines to minimize medical and other hazards that might arise from the environmental release of genetically engineered microorganisms (UNIDO 1989).

Collateral Readings

1. Burdon, R. D. 1994. The role of biotechnology in forest tree breeding. *Forest Genetic Resources,* 22: 2-5.
2. Burdon, R. D. 1999. Risk-management issues for genetically engineered forest trees. *NewZealand Journal of Forest Research,* 29: 375-390.
3. Convention on Biological Diversity. 2000. Cartagena Protocol on Biosafety. Internet document: www.biodiv.org/biosafe/Protocol/Protocol.html
4. DiFazio, S.P., Leonardi, S., Cheng, S. & Strauss, S.H. 1999. Assessing potential risks of transgene escape from fiber plantations. *Gene Flow and Agriculture: Relevance for Transgenic Crops,* 72: 171-176.
5. Englemann, F. 1997. *In vitro* conservation methods. *In* J.A. Callow, B.V. Ford-Lloyd and H.J. Newbury, eds. *Biotechnology and plant genetic resources: conservation and use,* p. 119-161. Biotechnology in Agriculture Series No. 19. Wallingford, UK, CAB International.
6. Fan, S.H. & Grossnickle, S.C. 1999. Clonal variation in gas exchange and freezing tolerance development of interior spruce (*Picea glauca* (Moench) Voss x *P. engelmannii* Parry ex Engelm.) during antumn acclimatisation. *Scandinavian Journal of Forest Research,* 14:240-250.
7. ffrench-Constant, R. & Bowen, D. 1999. Photorhabus toxins: novel biological insecticides. *Current Opinion in Microbiology,* 2: 284-288.
8. Frewen, B.E., Chen, T.H.H., Howe, G., Davis, J., Rohde, A., Boerjan, W. & Bradshaw, H.D. Jr. 2000. QTL and candidate gene mapping of bud set and bud set and bud flush in *Populus. Genetics.* 154: 837-845.
9. FSC. 1999. Principle and criteria for forest stewardship. Forest Stewardship Council (FSC) Internet document: www.fsc-uk.demon.co.uk/PrinciplesCriteria.html
10. Guo, W.W. & Deng. X.X. 1999. Intertribal hexaploid somatic hybrid plants regeneration from electrofusion between diploids of *Citrus sinesis* and its sexually incompatible relative. *Clausen lansium. Theoretical and Applied Genetics,* 98:581-585.
11. Gutierrez, R.A., MacIntosh, G.C. & Green, P.J. 1999. Current perspectives on mRNA stability in plants: multiple levels and mechanisms of control. *Trends in Plant Science,* 4: 429-438.

12. Lapierre, C., Pollet, B., Petti-Conil, M., Toval, G., Romero, J., Pilate, G., Leple, J.C., Boerjan, W., Ferret, V., De Nadai, V. & Jouanin, L. 1999. Structural alterations of lignins in transgenic poplars with depressed cinnamyl alcohol dehydrogenase or caffeic acid O-methyltransferase activity have an opposite impact on the efficiency of industrial kraft pulping. *Plant Physiology,* 119: 153-162.

13. Leiss, W. 1999. The trouble with science: public controversy over genetically modified foods. Paper presented at the Eastern Regional Meetings of the Canadian Society of Plant Physiologists, Kingston, Ontario, Canada, 12 December 1999. Internet document: www.ucalgary.ca/~wleiss/news/trouble_with_science.htm

14. Merkle, S.A. & Dean, J.F.D. 2000. Forest biotechnology. *Current Opinion in Biotechnology,* 11: 298-302.

15. OECD, 2000. Biotechnology regulatory developments in OECD member countries. Organization for Economic Co-operation and Development (OECD) Internet document: www.oecd.org/ehs/country.htm

16. Polonenko, D.R. 1999. Challenges and issues in scaling commercial production of conifer somatic embryogenesis. In Vitro *Cellular Development and Biology – Plant,* 35: 299-302.

17. Ralph, J., Mackay, J.J., Hatfield, R.D., O'Malley, D.M., Whetten, R.W. & Sederoff, R.R. 1997. Abnormal lignin in a loblolly pine mutant. *Science,* 277: 235-239.

18. Roberds, J.H. & Bishir, J. 1997. Risk analysis in clonal forestry. *Canadian Journal of Forest Research,* 27: 425-432.

19. Rugh, C.L., Senecoff, J.F., Meagher, R.B. & Merkle, S.A. 1998. Development of transgenic yellow poplar for mercury phytoremediation. *Nature Biotechnology,* 10: 925-928.

20. Santos, M. de Miranda & Lewontin, R.C. 1997. Genetics, plant breeding and patents: conceptual contradictions and practical problems in protecting biological innovations. *IPGRI Plant Genetic Resources Newsletter*, 112: 1-8.

21. Skinner, J.S., Meilan, R., Brunner, A.M. & Strauss, S.H. 1999. Options for genetic engineering of floral sterility in forest trees. *In* S.M. Jain and S.C. Minocha, eds. *Molecular biology of woody plants.* Vol. 1. Dordrecht, the Netherlands, Kluwer Academic Publishers.

22. TGERC. 1999. TGERC highlights of 1998-1999. Tree Genetic Engineering Research Cooperative (TGERC) Internet document: www.fsl.orst.edu/tgerc/1998-99.htm

23. Tremblay, L., Levasseur, C. & Tremblay, F.M. 1999. Frequency of somoclonal variation in plants of black spruce (*Picea mariana,* Pinaceae) and white spruce (*P. glauca,* Pinanceae) derived from somatic embryogenesis and identification of some factors involved in genetic instability. *American Journal of Botany*, 86: 1373-1381.

24. Zuffa, L., Lin, D. & Payne, P. 1999. Oneway crossing barriers in some interspecific crosses of Aigeiros and Tacamahaca. *Forestry Chronicle,* 75: 833-836.

Glossary

β-replicase: It helps in replication of viral RNA.

Accession: Plant or seed sample, strain or population held in a genebank or breeding programme for conservation and use.

Acetosyringone: One of the phenolic compounds released by wounded plant cells which can stimulate the transcription of the *vir* region of the *Ti* plasmid in the initial stages of transfer of DNA from *Agrobacterium* to the plant cell.

Acoustic coupler: An attachment for a Modem that allows data to be transmitted and received *via* a telephone handset rather than making direct electrical connections to the telephone network.

Activator: A substance or physical agent that stimulates transcription of a specific gene or operon.

Active Genepool Collection: A collection of germplasm used for regereration, multiplication, distribution, chracterization and evaluation. Ideally, germplasm in the active collection should be maintained in sufficient quantity to be available on request. Active collection germplasm is commonly duplicated in a base collection and is often stored under medium to long-term storage.

Agrobacterium rhizogenes: Bacterium causative for the hairy root disease, a form of organized tumor similar to the crown gall disease. Tumors also contain bacterial DNA transferred and covalently integrated in the plant genome.

***Agrobacterium tumefaciens*:** The bacterium causing crown gall disease of plants, inducing tumors to form. Tumors may be cultured in vitro and are useful in the study of this disease. The Ti plasmid of A. *tumefaciens*is known to cause the disease and a small portion of it is used as a vector inthe genetic modification of higher plant cells. Novel DNA sequences are spliced into the plasmid DNA segment, the DNA is circularized and introduced into the cultured plant cells.

Agroinfection: A method for initiating viral infection by inoculating host plants with *Agrobacterium* containing viral-specific DNA within the T-DNA.

Allotetraploid: hybrids between two species or genera containing the sum of the complete diploid chromosome complements of each parent.

Amplification Fragment Length Polymorphism (AFLP): A variant DNA amplification product produced by DAF, PCR, or RAPD of different size.

Antisense construct: A plasmid that contains the coding regions of a gene placed adjacent to a promotor in such an orientation as to cause the gene to be transcribed from the DNA strand complementary to that from which RNA is normally transcribed.

Antisense DNA: (1) The sequence of chromosomal DNA that is transcribed. (2) A DNA sequence that is complementary to all or part of a functional RNA (mRNA).

Antisense RNA: A RNA sequence that is complementary to all or part functional RNA (mRNA).

Artificial seed: Encapsulated or coated somatic embryos (embryoids) that are planted and treated like seed.

Asymmetric Somatic hybridization/partial genome transfer: The production of viable and fertile hybrid plant by irradiating protoplasts from one parental line (the donor) in order to pulverize the chromosomes and facilitate the elimination of substantial amount of the donor genome. Thereafter, the irradiated donor genome may be transferred into the genome of the recipient by protoplast fusion which can be carried out using three different methods: chemicals, an electrical field or a UV laser microbeam. Partial genome transfer system has potential use for the transfer of molecularly uncharacterized traits and multigenetic characters.

Autonomous replicating sequence (ARS): Any cloned DNA sequences that initiates and supports extrachromosomal replication of a DNA molecule in a host cell; often used in yeast cells. Also called autonomous(ly) replicating segment, autonomously replicating sequence.

Autoradiography: A technique that captures the image formed in a photographic emulsion as a result of the emission of either light or radioactivity from a labeled component that is placed next to unexposed film.

Auxin: One of a large class of plant hormones (phytohormones) allegedly produced in the growing tips of stems and roots. It causes cell enlargement and elongation but in excised, cultured systems they may promote cell division.

Biotechnology: The industrial use of biological processes; as yeast fermentation for alcohol production or plant cell culture for extraction of secondary products. The scope of biotechnology has been dramatically increased through genetic engineering.

Cell hybridization: Formation of synkaryons, viable cell **hybrids** produced through cell fusion. they are identifiable by their increased chromosome number compared to the parent cells and the possession of characters found in one or another of the parental cells.

Cell line: Developmental history or descent, through cell division from a single original cell; as callus may constitute a group of cell, all descendants from a single cell plating. Numerous **cell lines** may be present in a culture. Any deviation in culture technique may favor one cell line over another.

Cell number: The absolute number or approximation of the number of cells per unit area of a culture or medium volume.

Cell selection: Selection within a group of genetically different cells; involves competition between cells often under some stress. Selection criteria may involve cell viability 1 a unique phenotype or biochemical activity; or another basis for choice. Select cells or cell lines are generally relocated to fresh medium for continued selection and often are exposed to an increased level of the stress agent. The final objective is usually to regenerate plants from those select cells, in hope that the plants will exhibit the traits selected for at the cellular level.

Chimera or chimaera: A plant or tissue composed of more than one kind of genetic tissue. In a periclinal chimera one genotype occurs in a superficial layer, covering a genetically different core. In a mericlinal chimera one genotype occurs in a localized part of the plant and the rest is occupied by another genotype. In a sectorial chimera two genotypes share distinct sectors of the plant. Chimeras can occur naturally; be established artificially by grafting; or developed through somatic mutation by chemicals such as colchicine, or by other means in plant tissue cultures.

Crown gall: A disease of plants in which tumors form. The causal agent is the bacterium, *Agrobacterium tumefaciens*. Gall tissue has been grown in vitro. A tumor inducing portion of the bacterial genome (the Ti plasmid) my be used experimentally as a genetic tool to incorporate (vector) genetic information into plant cells.

Cybrid: A cell or plant cytoplasmic hybrid (heteroplast) with the nucleus of one and chtoplasmic organelles of another, or of both cells or plants.

Cytochimera or chromosomal chimera: Cells within a tissue possessing different chromosomal numbers, as when a callus consists of diploid and polyploid regions of cells.

Cytoplasmic hybrid or cybrid: A cell or plant hybrid (heteroplast) with the nucleus of one and the cytoplasmic organells of another or of both cells or plants.

De novo: (Latin: *fromthe beginning, anew*). Arising, sometimes spontaneously, from unknown or very simple precursors.

DNA ligases: Enzymes which bind DNA fragments together

Enhancer element: These are short regions of DNA which have the effect of increasing the levels of transcription, but unlike promotors, they affect genes from chromosomal loci at a long distance.

Enzyme: Any one of a number of spoecilized proteins produced in living cells, that speed the rate of specific chemical reactions, even at very low concentrations (organic catalyst), but is not used up in the reaction.

Epigenetic gene control: This is the new concept. It describes that the control of gene expression in higher eukaryotic organisms is related to the methylation of cytosine residues in gene coding sequence.

Excise: To cut, sever or otherwise remove or extract an organ or a segment of tissue from a plant or plant part; as the surgical removal of shoot tips. The process is **excision.**

Genetic engineering or recombinant DNA technology: Technology involving man-made changes in the genetic constitution of cells (apart from selective breeding). This technology usually employs a vector (such as the Ti plasmid of *Agrobacterium tumefaciens*) for transferring useful genetic information from a donor organism into a cell or organism that does not possess it. Genetic engineering has mny potential uses.

Genetic selection: Selection of genes, cells (cell selection), clones, etc., by man within populations or between populations or species. The usual purpose is to alter a specific phenotypic character. Such selection usually results in differential success rates of the various genotypes, reflecting many variables including selection pressure and genetic variability in populations.

Genetic transformation: The transfer of extracellular DNA (genetic information) among and between species; as, for example, with the use of bacterial or viral vectors.

Genetic variation: Differences in individuals derived from the same genotype in distincton to differences caused by the environment. **Genetic variance** denotes the proportion of pohenotypic variance caused by differences in the genetic make-up of an individual.

Genome: The basic haploid chromosome set of an individual; the sum of its **genes.**

Genotoxic: Carcinogenic; toxic to the chromosomes.

Geotaxis: Plant orientation with respect to gravity.

Germplasm or germ plasm: 1. The reproductive body tissues distinct from somatic (nonreproductive tissues). The genetic material, basis of heredity of an organism, passed on through previous generations. **2.** An individual representing a type of species or culture that may bve held in a repository for agronomic, historic (or other) reasons.

Gyrase: It is Type-II topoisomerase that catalyzes conversion of relaxed DNA to a superhelical form and requires ATP.

In vitro: (Latin = *in glass*). Experimentation on organisms or portions thereof in glassware or culture; growing under artificial conditions as in tissue culture. The antonym of in vivo.

In vivo: (Latin = *in life*) Experimentation on organisms under natural conditions within intact living organisms. the antonym of in vitro.

Introns: Sequences in DNAs which are in fact transcribed into mRNAs but are subsequently split out and there fore form no permanent constituents of the mRNAs (only in eucaryotes)

Jiffy pots: A brand name for peat pots that are sometimes used for ex vitro transplantation.

Kinases: They catalyze the transfer of the α-phosphate of a nucleoside 5'-triphosphate to the 5'-OH terminus of a DNA or RNA molecule.

Kinin: The original class name for substances promoting cell-division to which the prefix *cyto* has been added **(cytokinins)** to distinguish them from **kinins** in animals systems.

Ligases: They catalyze formation of phosphodiester linkages between DNA chain segments.

Ligation: Direct linkage of nucleic acid fragments

Lysis: Cell rupture or destruction, as through enzymatic action.

Maternal inheritance: Inheritance controlled by extrachromosoml (cytoplasmic) hereditary determinants.

Methionine (Met, $C_5H_{11}NO_2S$ mw 149.21): An amino acid precursor in ethylene synthesis and occasionally added to plant tissue culture media.

Methylases: They catalyze transfer of methyl groups from donar *S*-adenosyl-L-methionine (AdoMet) to specific bases of acceptor DNA molecules.

Micropropagation: Refers to propagation in culture by axillary or adventitious means. It is a general term for vegetative (asexual) in vitro propagation. It sometimes refers specifically to axillary bud proliferation.

Mitotic index (MI): The ratio of nuclei undergoing mitosis (including prophase) to total nuclei. MI = (Number of nuclei in mitosis/Total number of nuclei examined in the sample) X 100. A MI of 0.3 means that 30% of cells in the population are observed in mitosis.

Nucleases: They catalyze hydrolysis of the phospho-diester bonds of polynucleotides.

Nucleo-cytoplasmic ratio: The ratio of cell nuclear to cytoplasmic volume. This ratio is elevated in meristematic cells and low in differentiated cells.

Operator: Regulatory section in a nucleic acid with binding capacity for a repressor protein for blocking the synthesis of mRNAs.

Parasexual hybridization: Refers to genetic recombination by means other than through fertilization of germ cells (parasexual) leading to hybrid cells or individuals; as in hybrid cells or plants derived from somatic cell fusion.

Periclinal chimera: Refers to a condition in which geno-typically or cytoplasmically different tissues are arranged in concentric layers.

Phenocopy: A non-hereditary phenotypic change that is environmentally induced, during a limited developmental phase of an organism, that mimics the effect of a known genetic mutation.

Plasmid: Annular nucleic acid in cells outside the central chromosome with a lower molecular weight than the central chromosome; bearer of nonessential additional information, e.g., sex factors (F factors) and resistance factors (R ractors), plasmids and viral nucleic acids with known functions such as resistance or sequences are denoted by trivial abbreviation derived from their function or origin, e.g., SV 40 for simian virus, p^{BR322} for a plasmid imparting resistance to antibiotics. (*For detailed information see separate chapter on Plasmid in this book*)

Poly(A) adding enzymes: They attach AMP residues, from ATP, to the 3', 5' phosphodiester bond.

Polyethylene glycol (PEG) or carbowax: A fusion-inducing agent (fusogen) for agglutinating protoplasts which is used in somatic hybridization studies. This compound is also sometimes used in media as a non-metabolite osmoticum and is available in various molecular weights, ranging from ca. 200 to 6,000.

Polymerases: They catalyze addition of nucleoside triphosphates to 3'-OH terminals of single strands of preexiting polynucleotides (primers) with release of pyrophosphate.

Probe DNA: A radioactively lebeled (usually ^{32}P) DNA molecule used to detect complementary-sequence nucleic acid molecules by molecular hybridization. To localize the probe DNA sequence and reveal the complementary hybridization sequence auto-radiography is often used.

Protein digest: The enzymatic hydrolysis of proteins to yield their building block components, amino acids and short chain peptides (chains consisting of two to several amino acids normally without enzymatic function).

Protein stability: Protein, in eukaryotic cells have very diverse half-lives ranging from a few minutes to several hours. The degradation of unnecessary proteins is dependent upon the small protein ubiquitin, which when conjugated to proteins leads to their rapid degradation.

Protoplast fusion: The coalescence of the plasmalemma and cytoplasm of two or more **protoplasts** in contact with one another. Initial adhesion is a random process but coalescence may be promoted in various ways (induced fusion). When adhesion occurs between adjacent protoplasts during enzymatic wall degradation or between freshly isolated protoplasts in the absence of a fusion agent, it is termed spontaneous fusion.

Pure line: All cell or individual members are homozygous for one or more characters or genes and will give rise to more cells or organisms with the character(s) under consideration.

Recombinant DNA: Genetic material with novel gene sequences produced by cross-overs, chromosome reassortment, by other natural means or through genetic engineering.

Repression: Altered gene expression resulting in the failure of a specific protein synthesis.

Repressor: Regulatory protein with binding capacity for operators.

Resistance transfer factor: A plasmid, present in certain types of bacteria such as E. coli, that can impart resistance to antibiotics in animals exposd to them.

Restriction endonucleases: Enzymes which specifically cleave DNAs at a definite information-bearing sequence of nucleotides and thereby permit an insertion of foreign sequences. Restriction endonucleases are named in genetic engineering by trivial cell, e.g., EcoRI from *Escherichia coli* R Y 13, HindII from *Haemophilus influenza* Hpall from *Haemophilus parainfluenzq* or psti from *providencia stuartii*

Reverse transcriptase: It catalyze transcription of RNA nucleotide sequence into complementary DNA (cDNA).

Sodium hypochlorite (NaOH mw 74.44): A frequently used plant tissue sterilant at 0.5-2% w/v. The usual source is dilute (10-21% v/v) commercial laundry bleach, sometimes with surfactant is common. Tissue damage may occur if contact with tissue is prolonged. Thorough washing with sterile water generally follows treatment.

Transcription: Conversion of a DNA (gene) information sequence into a mRNA sequence

Transduction: Introduction of foreign genetic material into a cell by means of a virus or, in the case of bacteria, by means of phages

Translation control: In certain cases, after a gene has been transcribed, the mRNA can fold back upon itself to prevent its translation. This process is known as attenuation.

Tumor inducing principle (TIP): The plasmid carried by *Agrobacterium tumefaciens*, the crown gall organism. Through incorporation into the host genome the host tissue is transformed into tumor tissue.

Unwinding enzymes: They catalyze separation of complementary strends of DNA; require ATP.

Vector: Plasmid or virus that after ligation with a foreign DNA is used to introduce new information into the cell.

Virus-tested or virus-free: A plant that appears healthy and repeatedly tests negatively for the presence of one or more identifiable viruses. Such a plant may then be used as a stock or donor plant (explant source) for propagation purposes, and may be certified as virus tested (certified virus tested). The term virus-free is incorrect in most cases, as such a plant may contain one or more viruses which have not been assayed.

Abbreviations Acronyms

AAASA	Association for the Advancement of Agricultural Science in Africa
AA-N	Action Aid-Nepal
AATG	Actino Aid The Gambia
ABA	Abscissic acid
ABS	Agricultural Bank of Sudan
ACMV	African cassava mosaic virus
ACORD	Agency for Cooperation and Research in Development
AIC	Agricultural Inpus Corporation (Nepal)
AIMV	Alfalfa mosaic virus
AMCEN	African Ministerial Conference on the Environment
ARD	Associates in Rural Development (US)
ASC	Agricultural Service Centre (Nepal)
ASS	African Seeds of Survival (Ethiopia)
AVRDC	Asian Vegetable Research and Development Centre, China
BAP	6-Benzylaminopurine
BARC	Bangladesh Agriculture Research Council
BAU	Bangladesh Agricultural University
BMV	Brome mosaic virus
CaMV	Cauliflower mosaic virus
CAT	Chloramphenicol amino transferase
CBN	The cassava biotechnology network
cDNA	Complementary DNA
CEAO	Communaute Economique de l'Afrique de l'Ouest
CEPGL	Communaute aaeconomique des Pays des Grands Lacs
CESA	Ecuadorian Centre for Agricultural Services
CGIAR	Consultative Group for International Agricultural Research
CIAT	International Centre for Tropical Agriculture
CIC	Centro Internazionaale Crocevia
CIDA	Canadian International Development Authority

CILSS	Permanent Interstate Committee for Drought Control in the Sahel
CIP	International Potato Centre
CITES	Convention on International Trade in Endangered Species of Wild Fauna and Flora (IUCN), Washington D.C.
CLADES	Latin American Consortium for Agroecology and Development
CLRV	Cherry leafroll virus
CMV	Cucumber mosaic virus
CP	Coat Protein
CSB	Community Seed Bank (Philippines)
CSSTD	Central Seed Science and Technology Division (Nepal)
CyRSV	Cymbidium ringspot virus
DAR	Department of Agricultural Research (The Gambia)
DAS	Department of Agricultural Services (The Gambia)
DOA	Department of Agriculture (Nepal)
EC	European Community
ECOWAS	Economic Community of West African States
ELC	Environmental Liaison Centre International
EMI	Embu-Meru-Isiolo project (Kenya)
ENDA	Environemtn and Development Activities, Zimbabwe
ENS	Empresa Nacional de Sementes (Mozambique)
EPSP	5-enol pyruvyl shikimate 3-phosphate synthase
ESC	Ethiopia Seed Croporation
F1	First filial generation hybrid
FAO	UN Food and Agriculture Organization
FFHC	Freedom From Hunger Campaign
FV	Farmer's variety
GA3	Gibberellic acid
GATT	General Agreement on Tariffs and Trade
GCMV	Grapevine chrome mosaic virus
GCU	The Gambia Cooperative Union
GDP	Gross Domestic Product
GEMS	Global Environment Monitoring System
GIS	Geographic Information System
GNP	Gross National Product
GPMB	Gambia Produce Marketing Board
GPSN	Niassa Seed Production Board (Mozambique)
GRAIN	Genetic Resource Action International
GRAIN	Genetic Resources Action International
GRID	Global Resource Information Database
GSM	Good Seed Mission (The Gambia)
GTZ	Deutsche Gesellschaft fur Technische Zusammenarbeit

GUS	β-glucuronidase
HYV	High Yielding Variety
IAA	Indole-3-acetic acid
IARC	International Agricultural Research Centre
IBA	Indole-3-butyric acid
ICDA	International Coalition for Development Action
ICRISAT	International Crops Research Institute for the Semi-Arid Tropics (India)
IDESSA	Institute des Savanes
IDRC	International Devleopment Research Centre, Canada
IFOAM	International Federation of Organic Agricultural Movements
IITA	International Institute of Tropical Agriculture (Nigeria)
IJO	International Jute Organization
ILCA	International Livestock Centre for Africa
INIAP	National Agricultural Research Institute (Ecuador)
IRAT	Institute for Research in Tropical Agriculture and Food Crops (France)
IRAZ	Institut de Recherche Agronomique et Zootechnique de la CEPGL
IRRI	International Rice Research Institute (The Philippines)
ISTA	International Seed Testing Association
IUCN	World Conservation Union (formerly International Union for the Conservation of Nature and Natural Resources)
IVO	Development Research Centre (The Netherlands)
JICA	Japan International Cooperation Agency
KHAP	Koshi Hills Agriculture Project (Nepal)
Kin	Kinetin
KMP	Peasant Movement of The Philippines
KSHP	Kebkabiya Smallholder's Project (Nepal)
LEC	Livelihood Enhancement Committee (The Philippines)
MARs	Matrix attachment/associated regions
MASIPAG	Farmers-Scientists Partnership for Agricultural Development (The Philippines)
MCC	Mennonite Central Committee
MIND	Mindoro Institute for Development Inc (The Philippines)
MNC	Multi-National Corporation
MONAP	Mozambique Nordic Agriculture Programme
MV	Modern Variety
NAA	Naphthalene acetic acid
Na-alginate	Sodium alginate.
NACGRAB	National Centre for Genetic Resources and Biotechnology
NARC	National Agricultural Research Centre

NEF	Near East Foundation
NGO	Non-Governmental Organization
NIB	Replicase
NIHORT	National Horticultural Research Institute of Nigeria
NOS	Nopaline synthase
NPK	Nitrogen-Phosphorus-Potassium fertilizer
NPT-II	Neomycin phosphotransferase
NRI	Natural Resources Institute (UK)
NSA	National Seed Administration (Sudan)
NSB	National Seed Board (Bangaldesh)
ODA	Overseas Development Administration (UK)
ODI	Overseas Development Institute (UK)
OECD	Organization for Economic Cooperation and Development
OFSP	On-Farm Seed Production Project (The Gambia and Senegal)
OP	Open-pollinated variety
ORSTOM	Office de la Recherche Scientifique et Technique d' Outer-Mer (France)
PAC	Pakhribas Agricultural Centre (Nepal)
PAI	Paran Agronomy Institute (Brazil)
PARE	Partnership in Agricultural Research and Extension (Bangladesh)
PBR	Plant Breeder's Rights
PEBV	pea early browning virus
PEG	polyethyleneglycol
PG	Polygalaturonase
PGR	Plant Genetic Resources
PGRC/E	Plant Genetic Resources Centre/Ethiopia
PLRV	Potato leafroll virus
PMRA	Pest Management Regulatory Agency
PPS	Private Producer-Sellers Project (Nepal)
PRSV	Papaya Ring-spot Potyvirus
PTGS	Post-Transcriptional Gene Silencing
PVS	Potato virus S
PVX	Potato virus X
PVY	Potato virus Y
RAFI	Rural Advancement Foundation International
RENAMO	Mozambique National Resistance
RSS	Research and Specialist Services
RSV	Rice stripe virus
SADC	Southern African Development Community
SADCC	Southern African Development Coordination Conference
SAM	Sahabat Alam Malaysia

SARs	Scaffold attachment regions
SCF	Save the Children Federation (US)
SEAN	Seed Entrepreneurs Association of Nepal
SEARICE	South East Asia Regional Institute for Community Education
SEMOC	Sementes de Mozambiuqe Lda.
SHMV	Sunn hemp mosaic virus
SIBAT	Spring of Science and Technology (The Philippines)
SPG	Seed Producer Group (Nepal)
SRGB	SADC Regional Genebank
STIP	Seed Technology and Improvement Program (Nepal)
STU	Seed Technology Unit (The Gambia)
TAV	Tomato aspermy virus
TBSV	Tomato bushy stunt virus
TCV	Turnip crinkle virus
TEV	Tobacco etch virus
TGMV	Tomato golden mosaic virus
TMV	Tobacco mosaic virus
TobRV	Tobacco ringspot virus
ToMV	Tomato mosaic virus
TRIPs	Trade Related Aspects of Intellectual Property Rights
TRV	Tobacco rattle virus
TSWV	Tomato spotted wilt virus
TYLCV	Tomato yellow leaf curl virus
UN	United Nations
UNCED	UN Conference in Environment and Development
UNESCO	Un Educational, Scientific and Cultural Organization
UPLB	University of The Philippines at Los Banos
USAID	United States Agency for International Development
USC	Unitarian Service Committee (Cananda)
VRC	Village Relief Committee (Sudan)
WCMC	World Conservation Monitoring Centre
WIPO	World Intellectual Property Organization
WRI	World Resources Institute
WTO	World Trade Organisation
WWF	Worldwide Fund for Nature

Subject Index

Studies in Biotechnology Series No. 1

TRANSGENIC PLANTS

R. Ranjan

1. Introduction
2. Identification and Isolation of Genes
3. Gene Transfer Methods
4. Foreign Gene Expression in Plant
5. Engineered Resistance Against Fungal Pathogens
6. Engineered Resistance Against Virus Diseases
7. Engineered Resistance Against Bacterial Pathogens
8. Engineered Resistance Against Pest
9. Engineered Resistance Against Herbicides
10. Improvement of *Rhizobium* Inoculants by Genetic Engineering
11. Plant Genetic Engineering on Foods and Nutrition
12. Safety and Public Acceptance of Transgenic Plants
13. References

Studies in Biotechnology Series No. 2

HAIRY ROOT CULTURE
AND SECONDARY METABOLITES PRODUCTION

Usha Mukundan / Himanshu G. Dawda / Seema Ratnaparkhi

1. Crown Gall and Hairy Root Syndrome
2. Establishment and Growth Characteristics of Hairy Roots
3. Secondary Metabolite Production by Hairy Root Cultures
4. Scaling up of Hairy Root Cultures and Future Prospects
5. References

Studies in Biotechnology Series No. 3

SYNTHETIC SEEDS FOR COMMERCIAL CROP PRODUCTION

U. Kumar

1. Discovery and Future Perspectives of Synthetic Seeds
2. Somatic Embryogenesis: Principles, concepts and Applications
3. Plant Tissue Culture Principles and Methodology
4. Production of Synthetic Seeds
5. Economics of Tissue Culture and Synthetic Seed
6. References

Studies in Biotechnology Series No. 4

GENETIC ENGINEERING AND ITS APPLICATIONS

P. Joshi

1. Introduction
2. Gene Organization and Expression
3. Enzymes in Genetic Engineering
4. Gene Cloning Vectors
5. Gene Isolation, Identification and Synthesis
6. Cloning of Specific Gene
7. Specific Gene Transfer
8. Expression of Induced Genes
9. Applications of Genetic Engineering
10. Perspectives
11. References

Studies in Biotechnology Series No. 5

TRANSGENIC PLANTS PROMISE OR DANGER

B. L. Kakralya / Ishita Ahuja

1. Biotechnology and Transgenic Plants
2. Genetic Engineering and Transgenic Plants
3. Gene Isolation and Identification
4. Techniques of Producing Transgenic Plants
5. Expression of Induced Genes
6. Gene Silencing and Antisense Technology
7. Transgenic Plants and Photosynthetic Carbon Metabolism
8. Scopes and Achievements
9. Recommendations and Priorities for Future Research
10. Potential Ecological Impact of Transgenic Virus-Resistance in Plants
11. Alarming Issues Related to Transgenic Plants
12. Transgenic Plants: Promise and Danger
13. Collateral Readings, Glossary, Abbreviations Acronyms and Subject index

Studies in Biotechnology Series No. 6

PLANT BIOTECHNOLOGY AND BIODIVERSITY CONSERVATION

U. Kumar / M. Asija

1. Plant Biodiversity : Principles and Conservation
2. Ways to Conserve Plant Biodiversity
3. The Role of Plant Genetic Resources in Conserving Biodiversity
4. Biotechnological Techniques for Biodiversity Conservation
5. Biodiversity and Germplasm Conservation
6. Synthetic Seeds and Biodiversity Conservation
7. Conservation of Fruit and Plantation Crops Through Biotechnology
8. Conservation of Vegetable Crops Through Biotechnology
9. Conservation of Spices and Condiments Through Biotechnology
10. Conservation of Flower Crops Through Biotechnology
11. Glossary
12. Selected Bibliography
13. Some Interesting World Wide Web Sites